Springer Undergraduate M

Advisory Board

Other books in this series

Peter J. Cameron

Sets, Logic and Categories

 Springer

Peter Cameron, BSc, MA, D.Phil
School of Mathematical Sciences, Queen Mary and Westfield College,
London E1 4NS, UK

Cover illustration elements reproduced by kind permission of:
Aptech Systems, Inc., Publishers of the GAUSS Mathematical and Statistical System, 23804 S.E. Kent-Kangley Road, Maple Valley, WA 98038, USA. Tel: (206) 432 - 7855 Fax (206) 432 - 7832 email: info@aptech.com URL: www.aptech.com
American Statistical Association: Chance Vol 8 No 1, 1995 article by KS and KW Heiner 'Tree Rings of the Northern Shawangunks' page 32 fig 2
Springer-Verlag: Mathematica in Education and Research Vol 4 Issue 3 1995 article by Roman E Maeder, Beatrice Amrhein and Oliver Gloor 'Illustrated Mathematics: Visualization of Mathematical Objects' page 9 fig 11, originally published as a CD ROM 'Illustrated Mathematics' by TELOS: ISBN 0-387-14222-3, german edition by Birkhauser: ISBN 3-7643-5100-4.
Mathematica in Education and Research Vol 4 Issue 3 1995 article by Richard J Gaylord and Kazume Nishidate 'Traffic Engineering with Cellular Automata' page 35 fig 2. Mathematica in Education and Research Vol 5 Issue 2 1996 article by Michael Trott 'The Implicitization of a Trefoil Knot' page 14.
Mathematica in Education and Research Vol 5 Issue 2 1996 article by Lee de Cola 'Coins, Trees, Bars and Bells: Simulation of the Binomial Process' page 19 fig 3. Mathematica in Education and Research Vol 5 Issue 2 1996 article by Richard Gaylord and Kazume Nishidate 'Contagious Spreading' page 33 fig 1. Mathematica in Education and Research Vol 5 Issue 2 1996 article by Joe Buhler and Stan Wagon 'Secrets of the Madelung Constant' page50 fig 1.

British Library Cataloguing in Publication Data
Cameron, Peter J. (Peter Jephson), 1947-
  Sets, logic and categories. - (Springer undergraduate
  mathematics series)
  1. Set theory 2. Logic, Symbolic and mathematical
  3. Categories (Mathematics) 4. Set theory - Problems,
  exercises, etc. 5. Logic, Symbolic and mathematical -
  Problems, exercises, etc. 6. Categories (Mathematics) -
  Problems, exercises, etc.
  I. Title
  511.3
ISBN 1852330562

Library of Congress Cataloging-in-Publication Data
Cameron, Peter J. (Peter Jephson), 1947-
  Sets, logic and categories. / Peter J. Cameron.
    p.   cm. -- (Springer undergraduate mathematics series)
  Includes bibliographical references and index.
  ISBN 1-85233-056-2 (pbk. : alk. paper)
  1. Set theory 2. Logic, Symbolic and mathematical
3. Categories (Mathematics) I. Title. II. Series.
QA248.C17 1999   98-29309
511.3'22--dc21   CIP

Springer Undergraduate Mathematics Series ISSN 1615-2085
ISBN 1-85233-056-2
Springer Science+Business Media
springeronline.com

© Springer-Verlag London Limited 1998
4th printing 2005

Typesetting: Camera ready by author
Printed and bound at the Athenæum Press Ltd., Gateshead, Tyne & Wear
12/3830-543 Printed on acid-free paper SPIN 11401223

Preface

Set theory has a dual role in mathematics. In pure mathematics, it is the place where questions about infinity are studied. Although this is a fascinating study of permanent interest, it does not account for the importance of set theory in applied areas. There the importance stems from the fact that set theory provides an incredibly versatile toolbox for building mathematical models of various phenomena.

> Jon Barwise and Lawrence Moss, *Vicious Circles: On the Mathematics of Non-Wellfounded Phenomena* [4]

Reasoning and logic are to each other as health is to medicine, or – better – as conduct is to morality. Reasoning refers to a gamut of natural thought processes in the everyday world. Logic is how we ought to think if objective truth is our goal – and the everyday world is very little concerned with objective truth. Logic is the science of the justification of conclusions we have reached by natural reasoning.

> Julian Jaynes, *The Origin of Consciousness in the Breakdown of the Bicameral Mind* [24]

Much of Mathematics is dynamic, in that it deals with morphisms of an object into another object of the same kind. Such morphisms (like functions) form categories, and so the approach via categories fits well with the objective of organizing and understanding Mathematics. That, in truth, should be the goal of a proper philosophy of Mathematics.

> Saunders MacLane, *Mathematics: Form and Function* [36]

The three subjects in the title of this book all play a dual role in mathematics, which is hinted at in the quotations above. On the one hand, they are usually regarded as part of the foundations on which the structure of mathematics is built. This is obviously very important, philosophically as well as mathematically, since mathematics is commonly regarded as the most securely founded intellectual discipline of all. On the other hand, they are branches of mathematics in their own right: we use standard mathematical techniques to prove theorems in set theory, logic and category theory, and we use results from these areas in other parts of mathematics.

Here is one example of this, chosen from many possible. The Four-Colour Theorem asserts that any map drawn in the Euclidean plane (with reasonable assumptions about the shapes and borders of the countries) can be coloured with four colours in such a way that neighbouring countries are given different colours. This theorem was proved by Appel and Haken [2] with the aid of a very substantial computer calculation. Of course, the computer can only prove the result for *finite maps* (those involving only finitely many countries). But applying the Compactness Theorem of logic, it is a simple matter to deduce the infinite version from the finite.

We now give a brief sketch of what each of the three topics comprises.

Formal logic has two aspects, syntactic and semantic. The syntactic aspect is concerned with explaining which strings of symbols are to be regarded as formulae of a particular system, and which strings are 'theorems'. No meaning is attached to the symbols: the rules for testing whether a string is a well-formed formula, and the rules for manipulating formulae to prove 'theorems', are purely formal, and could be carried out by a computer without any intelligence. The semantic aspect is concerned with attaching meaning to the formulae, so that any formula expresses a mathematical fact (which may be true or false, or sometimes true and sometimes false depending on the values of variables it contains), and any theorem expresses a true fact about the systems to which it applies. We will deal with *first-order logic*, which is close to the actual practice of mathematicians: the structures to which it applies are sets on which various relations and functions are defined (including groups, ordered sets, and so on). We are concerned with the classes of structures satisfying particular formulae, or the theories (sets of formulae) holding in particular structures.

Set theory is the traditional foundation of mathematics. As the last paragraph suggests, the structures which first-order logic describes are based on sets. We are all familiar with definitions like 'A group is a set with a binary operation satisfying ...'. We will indicate, without giving a detailed proof, that anything in mathematics (from natural numbers to probability measures) can be interpreted as a set. On the other hand, set theory is a mathematical subject, and can be interpreted in different ways: different models of set theory support different mathematics. The best-known example of this is the Axiom of Choice, which has important applications in all branches of mathematics.

A more radical re-formulation of the basics is provided by category theory. This begins by emphasizing that it is the transformations between objects, rather than the objects themselves, which are really fundamental in mathematics. If two groups are isomorphic, we do not care that one is a group of matrices and the other a group of permutations. The transformations between structures can be pictured as arrows connecting various dots in a large diagram encompassing the branch of mathematics in question. What we need to know is how these transformations combine. It is possible to lay down rules for this. When this is done, we observe that certain mathematical objects (such as groups) themselves obey these axioms, giving a two-level structure to the subject. It is possible to build on this by considering the category of categories. Moreover, set theory is not really necessary to this foundation; sets and mappings form a category on the same footing as any other.

This book doesn't begin right at the beginning. It is assumed that you have met basic properties of sets (union, intersection, Venn diagrams, equivalence and order, one-to-one and onto functions), and the number systems (the natural numbers, integers, rational, real and complex numbers). This can be found in an introductory course on pure mathematics or discrete mathematics, such as Geoff Smith's book [42] in the SUMS series. It will help if you have met some logic before, but this is not vital. In a sense, this book follows on from David Johnson's book [26] in the series, although there are some differences in notation. Since examples of categories are taken from every branch of mathematics, the more you know the better; but you should at least have met familiar algebraic structures such as groups, rings and vector spaces from an axiomatic approach. This can be found in David Wallace's book [46] for abstract algebra, or Blyth and Robertson [7] for linear algebra.

Chapter 1 is about sets, from the point of view that we know what a set or collection of elements is. Right from the outset, we see that care is required to avoid such difficulties as Russell's Paradox. One of the main successes of set theory is to extend the theory of counting to infinite sets. In Chapter 2, we develop the theory of ordinal numbers, which 'count' certain kinds of ordered sets (*well-ordered sets*).

The next three chapters are about logic. Chapter 3 introduces the concept of a formal system, and treats propositional logic in detail; we see how the formal deduction and the semantics meet in the Soundness and Completeness Theorem for this logic. In Chapter 4, we do the same job for first-order logic, which treats the objects of mathematics (sets, relations and functions). Chapter 5 takes a few steps into model theory, which is concerned with the relation between the structures of mathematics and the logical formulae they satisfy.

In Chapter 6, we return to set theory, armed with the ideas of first-order logic, and give the Zermelo–Fraenkel axioms. One of these axioms is the (somewhat unintuitive) Axiom of Choice; we examine its consequences for mathematics. We also look at what can be said about classes which are not sets.

Chapter 7 is an account of category theory, the approach which treats functions rather than sets as basic. It is not yet possible to build mathematics on the foundation of category theory rather than set theory, but we look briefly at what can be done.

Finally, we take a very brief look at the philosophy of mathematics, and give some suggestions for further reading.

There is a World Wide Web site associated with this book, at the URL

<code>http://www.maths.qmw.ac.uk/~pjc/slc/</code>

This will contain solutions to all the exercises, a list of corrections, and links to related sites of interest on the Web.

The picture on p. 16 is by Neill Cameron.

Contents

1. Naïve set theory 1
   1.1 Handle with care 2
   1.2 Basic definitions 4
   1.3 Cartesian products, relations and functions 7
   1.4 Equivalence and order 11
   1.5 Bijections 15
   1.6 Finite sets 20
   1.7 Countable sets 24
   1.8 The number systems 28
   1.9 Shoes and socks 31

2. Ordinal numbers 37
   2.1 Well-order and induction 38
   2.2 The ordinals 39
   2.3 The hierarchy of sets 47
   2.4 Ordinal arithmetic 49

3. Logic 55
   3.1 Formal logic 56
   3.2 Propositional logic 58
   3.3 Soundness and completeness 64
   3.4 Boolean algebra 69

4.   First-order logic 79
   4.1   Language and syntax 80
   4.2   Semantics 83
   4.3   Deduction 85
   4.4   Soundness and completeness 88

5.   Model theory 95
   5.1   Compactness and Löwenheim–Skolem 95
   5.2   Categoricity 98
   5.3   Peano arithmetic 101
   5.4   Consistency 109

6.   Axiomatic set theory 113
   6.1   Axioms for set theory 114
   6.2   The Axiom of Choice 118
   6.3   Cardinals 124
   6.4   Inaccessibility 130
   6.5   Alternative set theories 133
   6.6   The Skolem Paradox 134
   6.7   Classes 136

7.   Categories 141
   7.1   Categories 143
   7.2   Foundations 146
   7.3   Functors 148
   7.4   Natural transformations 150

8.   Where to from here? 155
   8.1   Philosophy of mathematics 155
   8.2   Further reading 158

Solutions to selected exercises 161

References 175

Index 177

1
Naïve set theory

set, *n.*: aggregate, array, association, bale, batch, battery, body, bracket, bunch, bundle, cast, category, class, clump, cluster, collection, conglomerate, ensemble, family, genre, genus, group, grouping, ilk, kind, lot, mould, nature, order, pack, parcel, quantity, ring, section, sector, sort, species, style, type, variety.

Set theory is the most fundamental part of mathematics. The definition of almost any kind of mathematical object (a group, a ring, a vector space, a topological space, a Hilbert space ...) begins: a ⟨*thing*⟩ consists of a set, together with some extra structure in the form of operations, relations, subsets, sets of subsets, functions to the real numbers, or whatever. Also, as we will see, these operations, relations, etc. are themselves special kinds of sets.

However, it is very difficult to say what a set is. In the beginning, a set was simply a collection, class, or aggregate of objects, put together according to any rule we could imagine, or no rule at all. Of course this doesn't give a *definition* of a set, since if we try to explain what a collection, class or aggregate is, we find ourselves going round in circles. Later, we will see how this view is modified in the axiomatic approach, and the strengths and weaknesses of this approach.

In this chapter, after looking at Russell's Paradox as a cautionary tale, we review some of the basic notation and terminology about sets: subsets, unions and intersections, power sets, relations, functions, equivalence and order, finite and countable sets. The chapter ends by returning to Russell for another cautionary tale about the *Axiom of Choice*.

1.1 Handle with care

The foundations of mathematics have never been free of controversy. Set theory was developed by Georg Cantor in the late nineteenth century. But it soon became clear that the simple viewpoint that sets are just collections of elements, perhaps gathered together according to some rule, is untenable. *Russell's Paradox* demonstrates this. In fact, the logician Gottlob Frege was the first to develop mathematics on the foundation of set theory. He learned of Russell's Paradox while his work was in press, and wrote,

> A scientist can hardly meet with anything more undesirable than to have the foundation give way just as the work is finished. In this position I was put by a letter from Mr Bertrand Russell as the work was nearly through the press.

What then was this devastating communication?

It can be put in several related forms, of which the original is still the clearest. If arbitrary collections of objects form sets, we have to face the possibility that a set may be an element of itself: for example, the set consisting of all sets has this property. Most sets are not members of themselves, however. We may feel that it is slightly dangerous to have sets belonging to themselves, and restrict attention to 'good' sets which do not. Now Russell asked:

> *Let S be the set of all sets which are not members of themselves. Is S a member of itself?*

Either S is a member of itself, or it is not. Now, if S is a member of itself, then it is one of those sets which are not members of themselves, and so it is not a member of itself, which is a contradiction. On the other hand, if S is not a member of itself, then it satisfies the criterion of membership of S, which is also contradictory.[1]

Of other forms of this self-referential paradox, here are three.

One of the most famous is the *Liar Paradox*, invented by the Cretan philosopher and seer Epimenides. It is referred to by a number of classical authors, including the writer of the Pauline letter to Titus in the Christian Bible:

---

[1] Russell might have been pre-empted by the mediaeval theologian Thierry of Chartres, who postulated the 'Form of all other forms', which he interpreted as God. (Forms, in the Platonic sense, have something in common with the modern concept of sets.) His views were regarded as pantheistic and potentially heretical, but not as containing the seeds of a logical contradiction. See David Knowles, *The Evolution of Mediaeval Thought* [30].

> One of themselves [the Cretans], even a prophet of their own, said,
> the Cretans are alway liars, evil beasts, slow bellies.
>
> *Titus* 1, 12

If Epimenides said, 'All Cretans are liars', is his statement true or false?

An English adjective is called *autological* if it describes itself, and *hetero-logical* if it does not. So, for example, 'short' and 'pentasyllabic' are autological, while 'long' and 'trisyllabic' are heterological. Indeed, most adjectives are heterological: consider 'hard', 'soft', 'red', 'blue'. (What about adjectives like 'fluffy' or 'gnarled'?)

> *Is 'heterological' heterological?*

For the last paradox, we consider games between two players. We are not too precisely concerned with their structure; all that we require is that there are rules which determine whose move it is at any particular stage and what moves are allowed. When no move is possible, the game is over, though we allow the possibility that it ends in a draw.

A game in the above sense is called *well-founded* if any play of the game ends after finitely many moves (though there is not required to be a fixed upper bound for the number of moves). A trivial example is the following: the first player chooses a positive integer, and then the players take turns choosing positive integers, each smaller than the last one chosen. The game ends when no further choice is possible (that is, when the last number chosen is 1).

The *Hypergame* is played as follows. On the first move, the first player chooses any well-founded game. Then the players play that game, but with their roles swapped (so that the second player moves first in the chosen game). For example, if the first player chooses chess, then the second player takes the white pieces.[2]

Is the Hypergame well-founded? Obviously it is, since the games which can be chosen at the first move are required to be well-founded, and then the play of the Hypergame lasts only one move longer than the play of the chosen game.

But, since the Hypergame is well-founded, the first player may choose it on the first move. Then the second player starts playing the Hypergame by choosing a well-founded game. This player may also choose the Hypergame. Then the two players may continue choosing the Hypergame for ever, which contradicts the fact established above, that the Hypergame is well-founded.

---

[2] The rule that a game is drawn if the same position occurs for the third time guarantees that chess is well-founded. Littlewood [34] gives an upper bound of about $10^{10^{70.5}}$ for the number of possible chess games, comparable to the odds against a snowball surviving in Hell for a week.

Because of these paradoxes, the foundational subject of Set Theory must be set up with very great care. On the other hand, because of its importance, we must make the effort.

The matter is further complicated by *Gödel's Second Incompleteness Theorem*, an important result proved by Kurt Gödel in 1930. According to this theorem, we can never be sure that our set theory does not contain a contradiction lurking in its structure. Any formal theory which is strong enough to describe the natural numbers with their usual arithmetic, according to Gödel, cannot prove its own consistency. There are only two options: we can give a *relative consistency* proof, by proving the consistency of one subject within another. (This is what was done for non-Euclidean geometry, where models were constructed within Euclidean geometry.) Alternatively, we may simply work with the axiomatic theory and hope that any contradictions will come to light after a while. It is commonly accepted now that set theory does provide a secure foundation for mathematics.

In this chapter, we regard a set in the traditional way, but admit that not every collection of elements that can be imagined forms a set. Indeed, this is the positive conclusion we draw from Russell's Paradox:

Theorem 1.1

There is no set S such that $x \in S$ if and only if $x \notin x$.

Proof

If such a set S exists, then $S \in S$ if and only if $S \notin S$, which is a contradiction. So no such S can exist.   □

We develop the notation and terminology of set theory, allowing constructions of sets only if they are 'limited' in some way so as to avoid this paradox. Ultimately this procedure is not satisfactory; we need hard and fast rules. This will be motivated and developed in the chapter on axiomatic set theory. We will go on to a few more special topics: countable sets, and the construction of the number systems.

1.2 Basic definitions

We write $x \in y$ for 'x is a member of y'. This *membership relation* is the basic relationship; when we come to the axioms, it will be the undefined relation which is their subject.

The relation of equality can be defined in terms of membership, by the *Principle of Extension*, first stated by Leibniz:

Two sets are equal if they have the same members.

That is, $x = y$ if and only if, for all elements z, we have $(z \in x) \Leftrightarrow (z \in y)$.

The *empty set* is a set with no members. We say 'the empty set' rather than 'an empty set' because the Principle of Extension guarantees that there can't be more than one:

Theorem 1.2

There is only one empty set.

Proof

Let e_1 and e_2 be empty sets. Then for all elements z, the statements $z \in e_1$ and $z \in e_2$ are both false; so these statements are logically equivalent, and $e_1 = e_2$ by the Principle of Extension. $\square$

We write $\varnothing$ for the empty set.

One important consequence of what we have accepted so far is the doctrine that

> *Everything is a set.*

For, if a is an object that is not a set, then a has no members, and so a is equal to the empty set by the Principle of Extension, contradicting the assumption that it is not a set.

Given this doctrine, it is incumbent on us to show that it is reasonable, by constructing the everyday objects of mathematics (the number 27, the group of rotations of a regular dodecahedron, the Hilbert space $L^2[0,1]$, and so on) as sets. Of course, it would be much too tedious to do all this. But we will say enough in Section 1.8 to illustrate how such a programme could be carried out, and we will construct the natural numbers (as special cases of the *ordinal numbers*) as particular sets in Chapter 2.

It should be mentioned that there are other foundations of set theory which allow so-called *urelements*, basic elements which are not sets. In such theories, the Principle of Extension has to be modified so that it applies only to sets.

The set x is a *subset* of the set y (in symbols, $x \subseteq y$) if every member of x is a member of y, that is, if $(z \in x) \Rightarrow (z \in y)$ holds for all elements z.

Comparing this with the Principle of Extension, we see that $x = y$ if and only if both $x \subseteq y$ and $y \subseteq x$ hold.

We write $x \subset y$ to mean that x is a *proper subset* of y, that is, $x \subseteq y$ but $x \neq y$.

The *power set* of a set x (in symbols, $\mathcal{P}\,x$) is the set of all subsets of x. That is,

$$\mathcal{P}\,x = \{y : (\forall z)(z \in y) \Rightarrow (z \in x)\}.$$

As the name suggests, the power set of x is a set. However, if x is a large and complicated infinite set, the notion of a subset of x may not be entirely straightforward, and so there is a certain amount of vagueness in the notion of the power set of x. This vagueness will allow the possibility of several different versions of axiomatic set theory.

If $x_1, x_2, \ldots, x_n$ are finitely many sets, then we can gather them all into a set, which we denote $\{x_1, x_2, \ldots, x_n\}$. This is a set which has just these n elements as members.

Let x be a set; recall that the members of x are also sets. The *union* of x, written $\bigcup x$, is the set consisting of all members of the members of x: that is,

$$\bigcup x = \{z : z \in y \text{ for some } y \in x\}.$$

This notation is a bit unfamiliar. You will be more familiar with the union of two sets x_1 and x_2. This consists of all elements lying in either x_1 or x_2:

$$x_1 \cup x_2 = \{z : z \in x_1 \text{ or } z \in x_2\}.$$

It is just a special case of our general notion of union:

$$x_1 \cup x_2 = \bigcup\{x_1, x_2\}.$$

Similarly, the *intersection* of x, written $\bigcap x$, is the set consisting of all those elements which lie in every member of x: that is,

$$\bigcap x = \{z : z \in y \text{ for all } y \in x\}.$$

Thus, $\bigcap\{x_1, x_2\}$ is what is usually written as $x_1 \cap x_2$. We say that the sets x_1 and x_2 are *disjoint* if $\bigcap\{x_1, x_2\} = \varnothing$.

There is a problem with the empty set here (not the only place in mathematics where this occurs!). The union $\bigcup \varnothing$ is just the empty set, since there are no members of the empty set and hence no members of its members. But $\bigcap \varnothing$ should consist of everything in the universe: for, given any z, the condition that it belongs to every $y \in \varnothing$ is vacuously true (there is no $y \in \varnothing$ to provide a restriction). Because of Russell's Paradox, we do not want to allow this to be a set. Accordingly, we decree that $\bigcap x$ is defined only if $x \neq \varnothing$. (This is rather like the ban on dividing by zero in a field.) When we come to develop axiomatic set theory (in Chapter 6), we will have an axiom guaranteeing that unions exist, but not for intersections. The fact that intersections of non-empty sets exist will follow from the other axioms.

By contrast, there is no problem in considering the intersection of a set x which has the empty set as one of its members: this intersection is just the empty set!

We define the *difference* $X \setminus Y$ of two sets X and Y to consist of all elements which are in X but not in Y:

$$X \setminus Y = \{x \in X : x \notin Y\}.$$

1.3 Cartesian products, relations and functions

We next want to define the *ordered pair* (x_1, x_2). This should be an object constructed out of x_1 and x_2, with the property that

$$(x_1, x_2) = (y_1, y_2) \text{ if and only if } x_1 = x_2 \text{ and } y_1 = y_2.$$

The set $\{x_1, x_2\}$ does not have this property, since its elements do not come in any particular order: we have $\{x_1, x_2\} = \{x_2, x_1\}$. Some cleverness is needed to find the right construction. (But, once it is found, the details are not needed; only the property described above is ever used.)

We define

$$(x_1, x_2) = \{\{x_1\}, \{x_1, x_2\}\}.$$

Theorem 1.3

$$(x_1, x_2) = (y_1, y_2) \text{ if and only if } x_1 = y_1 \text{ and } x_2 = y_2.$$

Proof

The implication from right to left is clear. So suppose that $(x_1, x_2) = (y_1, y_2)$, that is, that

$$\{\{x_1\}, \{x_1, x_2\}\} = \{\{y_1\}, \{y_1, y_2\}\}.$$

The set on the left has just two members (which might happen to be equal), namely $\{x_1\}$ and $\{x_1, x_2\}$. Similarly for the set on the right. Thus the Principle of Extension shows that *either*

$$\{x_1\} = \{y_1\}, \quad \{x_1, x_2\} = \{y_1, y_2\} \quad (Case\ A),$$

or

$$\{x_1\} = \{y_1, y_2\}, \quad \{x_1, x_2\} = \{y_1\} \quad \textit{(Case B)}.$$

We consider these cases in turn.

In *Case A*, we have (again by Extension) $x_1 = y_1$, and either $x_1 = y_1$, $x_2 = y_2$ *(Case A1)*, or $x_2 = y_1$, $x_1 = y_2$ *(Case A2)*. In the first case, we have reached the conclusion we want. In the second, we have

$$y_2 = x_1 = y_1 = x_2,$$

so again we are done. Similarly, in *Case B*, we have $y_1 = x_1 = y_2$ and $x_1 = y_1 = x_2$, so again $x_1 = y_1$ and $x_2 = y_2$ follow.   □

Now we can define ordered n-tuples for $n > 2$ inductively by the rule

$$(x_1, \ldots, x_{n-1}, x_n) = ((x_1, \ldots, x_{n-1}), x_n).$$

They have the property that

$$(x_1, \ldots, x_n) = (y_1, \ldots, y_n) \text{ if and only if } x_1 = y_1, \ldots, x_n = y_n.$$

The *cartesian product* $X_1 \times X_2$ of two sets X_1 and X_2 is defined to be the set of all ordered pairs (x_1, x_2), where $x_1 \in X_1$ and $x_2 \in X_2$. We abbreviate $X \times X$ to X^2.

The term originates in the work of René Descartes. He realized that, by taking two perpendicular axes and setting up coordinates, the points of the Euclidean plane can be labelled in a unique way by ordered pairs of real numbers. We can then take the further step of saying that a point *is* a pair of real numbers, so that the set of points of the Euclidean plane is the cartesian product $\mathbb{R} \times \mathbb{R}$. Then we can say, for example, that a line or curve *is* the set of solutions of some equation in two variables, and we are well on the way to turning geometry into a branch of algebra. (Of course, this is a much less grandiose project than turning all of mathematics into set theory, but it is a step on this road!)

In a similar way, $X_1 \times \cdots \times X_n$ is the set of all ordered n-tuples $(x_1, \ldots, x_n)$, where $x_1 \in X_1, \ldots, x_n \in X_n$; and we abbreviate $X \times \cdots \times X$ (n factors) to X^n.

Note that, unlike the case of union and intersection, we have not defined a concept of the cartesian product $\prod x$ of an arbitrary set x, so that $\prod\{X_1, X_2\} = X_1 \times X_2$. This can be done (more or less), but we need some more notation to set it up.

The concept of a *function* has changed over the centuries. Until the nineteenth century, a function was given by a formula, though this formula could involve such analytic processes as sums of infinite series or transcendental functions as well as arithmetic operations. In Forsyth's *Theory of Functions of a Complex Variable*, written as late as 1893 and quoted by Littlewood [34], it is explained thus:

> ... if the value of X depends on that of x and on no other variable magnitude, it is customary to regard X as a function of x; and there is usually an implication that X is derived from x by some series of operations.

A crisis was provoked by various discoveries at the end of the nineteenth century, among them a real function which is everywhere continuous but nowhere differentiable, and a curve that passes through every point of the unit square. A more up-to-date version might involve specifying that the function values can be calculated by a computer, at least in principle. The modern definition however does not require this; it is as general as possible, subject only to the requirement that 'everything is a set': we ask only that each 'input' determines a unique 'output'.

A *function* $f : X \to Y$ from the set X to the set Y is a subset of $X \times Y$ with the property that *every element of X is the first component of a unique ordered pair in f;* in other words, for any $x \in X$, there is a unique $y \in Y$ such that $(x, y) \in f$. We write $y = f(x)$ as an alternative to $(x, y) \in f$. This only defines the compound expression $y = f(x)$; but, of course, there is a natural way of giving a meaning to $f(x)$, as the unique element y of Y for which this equation holds.

The requirement that f is a function can be expressed formally as follows:

- $(\forall x \in X)(\exists y \in Y)((x, y) \in f)$;

- $(x, y_1), (x, y_2) \in f \Rightarrow (y_1 = y_2)$.

The function $f : X \to Y$ is said to be *injective* or *one-to-one* if distinct elements of X have distinct images – that is,

- $(x_1, y), (x_2, y) \in f \Rightarrow (x_1 = x_2)$;

it is *surjective* or *onto* if every element of Y is the image of some element of x – that is,

- $(\forall y \in Y)(\exists x \in X)((x, y) \in f)$;

and it is *bijective* if both of these conditions hold. A bijective function has the effect of 'matching up' the elements of X with those of Y, so that there are the 'same number' of elements in X and Y. (This remark will become significant when we come to investigate exactly what numbers to use to measure the size of a set.)

A bijective function $f : X \to Y$ has an *inverse* $f^{-1} : Y \to X$, defined by

$$f^{-1} = \{(y, x) : (x, y) \in f\}.$$

Two functions $f : X \to Y$ and $g : Y \to Z$ can be *composed* to give a function $f \circ g : X \to Z$ defined by

$$f \circ g = \{(x, z) : (\exists y \in Y)((x, y) \in f \text{ and } (y, z) \in g)\}.$$

(Note that $(f \circ g)(x) = g(f(x))$: this unfortunate reversal arises because of our notation for functions, writing the function on the left of its argument.)

A particular bijection is the *identity* function i_X on a set X, which is defined to be

$$i_X = \{(x, x) : x \in X\}.$$

Note that, if $f : X \to Y$ is any function, then $i_X \circ f = f \circ i_Y = f$; and, if f is a bijection, then $f \circ f^{-1} = i_X$, and $f^{-1} \circ f = i_Y$.

Sometimes we use the notation Y^X for the set of all functions from X to Y.

A *relation* between X and Y is just a subset of $X \times Y$. If $X = Y$, we speak of a *binary relation* on X. The reason is much as before: we want to describe relations like 'less than' or 'divides', but without making any demands on the relation other than that it is a set. We know everything about a relation when we know which ordered pairs satisfy it. Note that a function is a special kind of relation.

If R is a relation, we often write $x \, R \, y$ instead of $(x, y) \in R$. This is motivated by the common relations such as 'less than' or 'divides', which are written $x < y$ and $x \mid y$ respectively. However, even though we regard $<$ as a set of ordered pairs, we draw the line at writing expressions like $(x, y) \in <$, however correct they may be logically!

Any set has the property that all its members are sets. However, we need a more general concept, that of a *family of sets*, which is more like an ordered tuple, and has the property that its components can be repeated more than once.

A *family of sets*, indexed by the *index set* I, is a function $F : I \to W$ for some W. The sets in the family are the sets $X_i = F(i)$ for $i \in I$, and we often write the family as $(X_i : i \in I)$. (Note the round brackets, suggesting an ordered tuple!)

The *union* and *intersection* of a family of sets are defined by

$$\bigcup_{i \in I} X_i = \bigcup \{X_i : i \in I\},$$
$$\bigcap_{i \in I} X_i = \bigcap \{X_i : i \in I\}.$$

(This means, for example, that $\bigcup_{i \in I} X_i$ consists of all sets which are members of X_i for some $i \in I$.) The union is always defined, and the intersection is defined provided that $I \neq \varnothing$.

We can also define the cartesian product of a family of sets. First, given a family $(X_i : i \in I)$, defined by a function $F : I \to W$ (so that $X_i = F(i)$), we define a *choice function* for the family to be a function $f : I \to \bigcup W$ having the property that $f(i) \in F(i)$ for all $i \in I$. (The function f 'chooses' a representative element from each set in the family.) The choice function is itself a family; we often write it as $(x_i : i \in I)$, with $x_i = f(i)$. Now the *cartesian product* $\prod_{i \in I} X_i$ is defined to be the set of all choice functions for the family $(X_i : i \in I)$.

For example, suppose that $I = \{1, 2\}$, and let $(X_i : i \in I)$ be a family of sets indexed by I. A choice function picks out an element $x_1 \in X_1$ and an element $x_2 \in X_2$, and so can be represented as an ordered pair (x_1, x_2), an element of $X_1 \times X_2$. So $\prod_{i \in I} X_i$ is essentially the same as $X_1 \times X_2$ in this case.

If one of the sets in the family is empty, then the cartesian product of the family is empty: no choice function can exist. Is the converse true? It turns out that the converse is the celebrated *Axiom of Choice*, about which we will have more to say later. We note here that it can neither be proved nor disproved by the reasonable assumptions about sets that we have made so far.

For future record, we display it here, not as an 'axiom' (yet), far less as a 'theorem', but as a mysterious principle we will invoke from time to time.

Axiom of Choice: Any family of non-empty sets has a choice function.

1.4 Equivalence and order

Some special kinds of relations will be very important to us. To define these, we first list some properties which a relation may (or may not) possess.

The relation R on X is said to be

- *reflexive* if $(x, x) \in R$ for all $x \in X$;

- *irreflexive* if $(x, x) \notin R$ for all $x \in X$;

- *symmetric* if $(x, y) \in R$ implies $(y, x) \in R$;

- *antisymmetric* if $(x, y) \in R$ and $(y, x) \in R$ imply $x = y$;

- *transitive* if $(x, y) \in R$ and $(y, z) \in R$ imply $(x, z) \in R$.

Note that 'irreflexive' is not the same as 'not reflexive', and 'antisymmetric' is not the same as 'not symmetric'. Moreover, we do not define 'antisymmetric' to mean that $(x, y) \in R$ implies $(y, x) \notin R$: for this would mean that an antisymmetric relation would automatically be irreflexive, and we don't want to prejudge this.

An *equivalence relation* on X is a relation which is reflexive, symmetric and transitive. As is well known, the job that an equivalence relation performs is to partition a set. We now make the appropriate definitions to formulate this property.

If R is any relation on X, we define the *R-class* of an element x to be the set $\{y \in X : (x, y) \in R\}$ consisting of everything related to x.

A *partition* of X is a set of non-empty sets which cover X without overlap. More precisely, it is a set P with the properties

- for all $p \in P$, $p \neq \varnothing$;

- for all $p, q \in P$, $p \cap q = \varnothing$;

- $\bigcup P = X$.

(Note how simply the third condition can be stated using our notation for union.)

Theorem 1.4 (Equivalence Relation Theorem)

(a) If R is an equivalence relation on X, then the set

$$P = \{R(x) : x \in X\}$$

of R-classes is a partition of X.

(b) If P is a partition of X, then

$$R = \{(x, y) : x, y \in p \text{ for some } p \in P\}$$

is an equivalence relation on X.

(c) The constructions in (a) and (b) are mutually inverse.

The last statement means that, if we start with an equivalence relation, construct a partition as in (a), and then construct the equivalence relation as in (b), then it is equal to the original equivalence relation; and similarly if we start with a partition. This theorem is a fundamental mathematical statement, and a proof will not be provided here since it is standard. See Smith [42], Proposition 1.4; Johnson [26], Theorem 4.1; Wallace [46], Theorem 1.8.

If R is an equivalence relation on X, we use the notation X/R for the set of equivalence classes (the partition of X corresponding to R in the Equivalence

Relation Theorem). The notation is chosen to resemble that for a factor group or factor ring; deliberately, as we now see.

Suppose that X and Y are sets, and f a function from X to Y. The *image* of f, written $\mathrm{Im}(f)$, is the set

$$\mathrm{Im}(f) = \{y \in Y : (\exists x \in X)(y = f(x))\}$$

of elements which can be represented as $f(x)$ for some $x \in X$. The *kernel* of f, written $\mathrm{KER}(f)$, is the relation on X defined by

$$\mathrm{KER}(f) = \{(x_1, x_2) \in X \times X : f(x_1) = f(x_2)\}.$$

Theorem 1.5 (First Isomorphism Theorem)

Let f be a function from X to Y. Then

(a) $\mathrm{Im}(f)$ is a subset of Y;

(b) $\mathrm{KER}(f)$ is an equivalence relation on X;

(c) f induces a bijection from $X/\mathrm{KER}(f)$ to $\mathrm{Im}(f)$.

The proof is an exercise.

There are two kinds of order relations, modelled by the relations 'less than' or 'less than or equal' on the integers or real numbers. They are defined as follows. A *(strict) (partial) order* on X is a relation on X which is irreflexive, antisymmetric and transitive. A *non-strict (partial) order* on X is a relation which is reflexive, antisymmetric and transitive. Note the use of brackets: it is intended that the term 'strict' is the default, so that if we don't specify whether an order is strict or not it is assumed to be strict; and the term 'partial' is the default, so that (when we shortly meet the term 'total order') it is assumed that an order is partial unless it is stated that it is total.

So, on one of the standard number systems, the relation $<$ is a strict order, while $\leq$ is a non-strict order.

An example of a non-strict order is the relation $\subseteq$ or 'is a subset of', on any set x. (The end of this sentence is a little surprising: you may have expected the statement 'on the power set $\mathcal{P}x$ of any set x'. But remember that everything is a set!)

Often we denote strict and non-strict orders by $<$ and $\leq$ respectively. There is a close relationship between the two types:

Theorem 1.6 (Order Theorem)

(a) If R is a strict order on X, then

$$S = R \cup \{(x,x) : x \in X\}$$

is a non-strict order on X.

(b) If S is a non-strict order on X, then

$$R = \{(x,y) \in S : x \neq y\}$$

is a strict order on X.

(c) The constructions in (a) and (b) are mutually inverse.

Again we omit the straightforward proof. We adopt the convention that if a strict and a non-strict order are related as in the Order Theorem, and the strict order is called $<$, then the non-strict order is called $\leq$, and *vice versa*. Also, we write $x > y$ as an equivalent of $y < x$, and $x \geq y$ as an equivalent of $y \leq x$. As a final piece of terminology, if $<$ is an order on X, we call the ordered pair $(X, <)$ an *ordered set*.

A *total order* is a partial order which satisfies the condition of *trichotomy*: for all $x, y \in X$, one of $(x,y) \in R$, $x = y$, and $(y,x) \in R$. Note that, if R is a strict order, then at most one of these three conditions can hold for any x and y, so R is total if exactly one always holds. Also, if we write $x < y$ for $(x,y) \in R$, the condition of trichotomy can be written: for all $x, y \in X$, one of $x < y$, $x = y$, $x > y$ holds.

Let $(X, <)$ be an ordered set. We call the element $x \in X$ a *least element* of X if, for all $y \in X$, we have $x \leq y$. We call x a *minimal element* of X if, for all $y \in X$, $y \leq x$ implies $y = x$. (So 'least' means 'smaller than everything else'; and 'minimal' means 'nothing else is smaller'.)

Theorem 1.7

Let $(X, <)$ be an ordered set.

(a) If a least element of X exists, then it is minimal, and moreover it is the unique minimal element.

(b) If X is totally ordered, then any minimal element is a least element.

Proof

(a) Suppose that x is least. We must show first that it is minimal. So suppose that $y \leq x$. Then also $x \leq y$, since x is least; so $x = y$ by antisymmetry.

Now let y be any minimal element. Then $x \le y$, since x is least; and so $x = y$, since y is minimal. So x is the unique minimal element.

(b) Suppose that X is totally ordered, and that x is minimal. If x is not least, then there exists $y \in X$ such that $x \le y$ does not hold. Hence $y < x$ by trichotomy. But this is impossible, since $y \le x$ implies $y = x$, as x is minimal.

$\square$

Dually, an element x of the ordered set $(X, <)$ is *greatest* if $x \ge y$ for all $y \in X$; and x is *maximal* if $y \ge x$ implies $y = x$. The analogues of the above theorem hold for these concepts.

Finally in this section, we define the concept of *isomorphism*. The definition is given only for sets with a single relation, but can be reworked to apply much more generally.

Let R be a relation on a set X, and S a relation on a set Y. We say that the structures (X, R) and (Y, S) are *isomorphic* if there is a bijection $f : X \to Y$ with the property that for all $x_1, x_2 \in X$, if $(x_1, x_2) \in R$, then $(f(x_1), f(x_2)) \in S$, and conversely. So the sets X and Y can be matched up so that the relation is satisfied by corresponding pairs in the two sets. We write $(X, R) \cong (Y, S)$ to denote that the structures (X, R) and (Y, S) are isomorphic. If the relations are understood, we sometimes abuse the notation and simply say that X and Y are isomorphic (and write $X \cong Y$).

Isomorphism is an 'equivalence relation' on the class of relational structures: that is,

- (X, R) is isomorphic to itself (by means of the identity function);

- if $(X, R) \cong (Y, S)$ (by a bijection f), then $(Y, S) \cong (X, R)$ (by the inverse bijection);

- if $(X, R) \cong (Y, S)$ (by the bijection f) and $(Y, S) \cong (Z, T)$ (by the bijection g), then $(X, R) \cong (Z, T)$ (by the composition of f and g).

1.5 Bijections

One of our goals is to develop a system of numbers to measure the sizes of sets, finite or infinite. The natural numbers play this role for finite sets. Though we haven't defined the natural numbers yet, it will turn out that the number n is a 'standard' set with n elements, and an arbitrary set has n elements if and only if it can be put into one-to-one correspondence with the 'standard' set n. In fact, the notion that two sets 'have the same number of elements' is really

Fig. 1.1. Counting sheep

more basic than any statement about what this number is, since it only requires that there is a bijection between the sets. (See Figure 1.1.) Georges Ifrah [23] tells the story of an American archaeological team working in the palace of Nuzi, near Kirkuk in modern Iraq. They found a clay envelope inscribed with a list of 48 sheep and goats; when opened, the envelope contained 48 clay balls. Presumably the clay envelope was made by a literate accountant, while the clay balls were to enable the shepherd to check that the flock was complete. The shepherd would simply have checked that there was a bijection between the clay balls and the sheep, and would not have needed to be able to count to 48. The significance of the find was brought home to the archaeologists when their uneducated servant, sent to the market to buy chickens, was unable to say how many chickens he had purchased, but produced a collection of pebbles, one for each chicken.

So we take the bold approach: we say that two sets X and Y *have the same cardinality* if there is a bijection between them – we do not define yet what the cardinality of a set is. We write $|X| = |Y|$ if X and Y have the same cardinality, but, again, we do not yet assign any meaning to the symbol $|X|$ in isolation. (This will be done later!)

More generally, we say that the set X *has smaller cardinality than* the set Y (in symbols, $|X| \leq |Y|$) if there is an injection (a one-to-one mapping) from X to Y. If this holds, and if X and Y do not have the same cardinality, then we say that X *has strictly smaller cardinality than* Y, and write $|X| < |Y|$.

Surprisingly, many assertions which might seem quite obvious or natural cannot be proved without the Axiom of Choice. These include the statements

- any two sets X and Y are *comparable* (in the sense that either $|X| \leq |Y|$ or

$|Y| \leq |X|$); and

- if $X \neq \varnothing$, there is an injective function from X to Y if and only if there is a surjective function from Y to X.

This being the case, it is important to see just what we can prove. At least the following two statements are true.

Theorem 1.8

If there is an injective function from X to Y, and $X \neq \varnothing$, then there is a surjective function from Y to X.

Proof

Let $f : X \to Y$ be injective. Let a be an arbitrary element of X. Now define a function $g : Y \to X$ by the rule

$$g(y) = \begin{cases} x & \text{if } f(x) = y; \\ a & \text{if no such } x \text{ exists.} \end{cases}$$

Since f is injective, if x exists, then it is unique; so the function is well-defined. Now for any $x \in X$, we have $x = g(f(x))$; so g is surjective.   □

Theorem 1.9 (Schröder–Bernstein Theorem)

If there is an injective function from X to Y and an injective function from Y to X, then there is a bijective function from X to Y.

In other words, if $|X| \leq |Y|$ and $|Y| \leq |X|$, then $|X| = |Y|$.

Proof

We are given injective functions $f : X \to Y$ and $g : Y \to X$, and have to construct from them a bijection between the two sets. We give two similar proofs of this important result. The first is more intuitive, but uses some elementary properties of natural numbers, whereas the second uses nothing but set theory. Without loss of generality, X and Y are disjoint. (Given any sets X and Y, we can find disjoint sets X' and Y' bijective with X and Y: for example, take $X' = X \times \{1\}$ and $Y' = Y \times \{2\}$.) This dodge is not needed for the second proof.

First Proof

We say that $y \in Y$ is the *parent* of $x \in X$ if $g(y) = x$; dually, $x \in X$ is the *parent* of $y \in Y$ if $f(x) = y$. Each element of X or Y has at most one parent. An *ancestral chain* for $z \in X \cup Y$ is a tuple $(z_0, z_1, \ldots, z_n)$ such that $z_0 = z$ and z_{i+1} is the parent of z_i for $i = 0, \ldots, n-1$. (Its elements belong alternately to X and Y.) The *length* of such a chain is n (the number of *steps*, not the number of *elements*).

Now there are two possibilities for any element z. Either there are arbitrarily long ancestral chains for z, in which case we shall say that z has *infinite depth*; or there is a unique longest ancestral chain for z, ending with an element with no parent, in which case we say that the length of this chain is the *depth* of z. (The second possibility includes the case when z itself has no parent, in which case its depth is 0.) We let X_e denote the set of elements of X whose depth is even; X_o the set of elements of X with odd depth; and X_∞ the set of elements with infinite depth. We define Y_e, Y_o, and Y_∞ similarly.

If $x \in X$ has finite depth, then $f(x)$ has depth one greater than the depth of X; and if $x \in X$ has infinite depth, then so does $f(x)$. So f maps $X_e \to Y_o$, $X_o \to Y_e$, and $X_\infty \to Y_\infty$. A similar assertion holds for the action of g on elements in Y. Furthermore, elements of Y_o or Y_∞ have parents; so f maps $X_e \to Y_o$ and $X_\infty \to Y_\infty$ bijectively. (This does not hold for $X_o \to Y_e$ since an element of Y_e may have no parent.)

Define a map $h : X \to Y$ by

$$h(x) = \begin{cases} f(x) & \text{if } x \in X_e \cup X_\infty, \\ g^{-1}(x) & \text{if } x \in X_o. \end{cases}$$

Then it is easily seen that h is a bijection.   □

Second Proof

We first prove a couple of lemmas.

Lemma 1.1

Let X be a set, and $p : \mathcal{P}X \to \mathcal{P}X$ a function which is monotonic, in the sense that if $A \subseteq B \subseteq X$, then $p(A) \subseteq p(B)$. Then there is a set $Z \subseteq X$ such that $p(Z) = Z$.

Proof

We set $Z = \bigcup\{A \subseteq X : A \subseteq p(A)\}$. Take $z \in Z$. Then there is a set $A \subseteq X$ such that $z \in A$ and $A \subseteq p(A)$. So $z \in p(A)$. Moreover, $A \subseteq Z$, so $p(A) \subseteq p(Z)$ by hypothesis. Thus $z \in p(Z)$. We have shown that $Z \subseteq p(Z)$. Again by hypothesis, $p(Z) \subseteq p(p(Z))$. So $p(Z)$ is one of the sets in the family whose union is Z. But this means that $p(Z) \subseteq Z$. So we have $Z = p(Z)$, as claimed.

$\square$

For the next lemma, we require some notation. If f is a function on a set X, and $A \subseteq X$, we define $f[A] = \{f(a) : a \in A\}$.

Lemma 1.2

Let X and Y be sets, and $f : X \to Y$ and $g : Y \to X$ be injective functions. For $A \subseteq X$, set $p(A) = X \setminus g[Y \setminus f[A]]$. Then, if $A \subseteq B \subseteq X$, we have $p(A) \subseteq p(B)$.

Proof

Suppose that $A \subseteq B$ but $p(A) \not\subseteq p(B)$. Choose an element $x \in p(A) \setminus p(B)$. Then, by definition of p, we have $x \in g[Y \setminus f[B]]$; say $x = g(y)$, where $y \in Y \setminus f[B]$. But $A \subseteq B$, so $f[A] \subseteq f[B]$, so $Y \setminus f[B] \subseteq Y \setminus f[A]$. We conclude that $y \in Y \setminus f[A]$, so that $x = g(y) \in g[Y \setminus f[A]]$; that is, $x \notin p(A)$, contrary to assumption.

$\square$

Now we prove the Schröder–Bernstein Theorem. Let X, Y, f, g be as in the theorem. Define the function $p : \mathcal{P} X \to \mathcal{P} X$ as in Lemma 1.2. Then the two lemmas imply the existence of a subset Z of X with $p(Z) = Z$. Define a function $h : X \to Y$ as in the first proof:

$$h(x) = \begin{cases} f(x) & \text{if } x \in Z, \\ g^{-1}(x) & \text{if } x \notin Z. \end{cases}$$

Note that, if $x \notin Z$, then $x \in g[Y \setminus f[Z]]$, so that $f = g(y)$ for some unique $y \in Y \setminus f[Z]$; this element y is what we have called $g^{-1}(x)$. Now it is readily checked that $h : X \to Y$ is a bijection.

$\square$

There is one particular case, noted by Cantor, where we can show that no bijection exists between two sets.

Theorem 1.10 (Cantor's Theorem)

For any set X, there is an injection from X to $\mathcal{P} X$ but no bijection between these sets; that is,

$$|X| < |\mathcal{P} X| .$$

Proof

We can define an injection $f : X \to \mathcal{P} X$ very simply: set $f(x) = \{x\}$. (By the Principle of Extension, if $\{x\} = \{y\}$ then $x = y$.)

For the second statement, we suppose for a contradiction that there is a bijection $h : X \to \mathcal{P} X$. Let

$$Y = \{x \in X : x \notin h(x)\}.$$

Since h is a bijection, there is a (unique) element $y \in X$ such that $Y = h(y)$. Now we ask: *Is $y \in h(y)$?* (Note the similarity with Russell's Paradox.) If $y \in h(y)$, then by definition $y \notin Y$; and similarly if $y \notin h(y)$ then $y \in Y$. But these contradict the fact that $Y = h(y)$. $\square$

1.6 Finite sets

A set has n elements if it has a bijection with the set $\{1, \ldots, n\}$. (This is the basic principle of counting, that we learn at an early age and use every day.) It will be convenient to regard the number n as a 'standard' n-element set to use for counting. However, if we take n to be the set $\{1, \ldots, n\}$, then n is a member of itself; a feeling of nervousness about Russell's Paradox leads us to proceed slightly differently. Anticipating the next chapter, we take the number n to be the set $\{0, \ldots, n-1\}$. Also, we take the set $\mathbb{N}$ of natural numbers to include 0: that is, $\mathbb{N} = \{0, 1, 2, \ldots\}$.

So set theorists count

zero, one, two, ...

instead of

one, two, three, ...

like the rest of us.

We begin with a simple but important property: no two natural numbers have the same cardinality!

Theorem 1.11

If there is a bijection between n and m, then $n = m$.

Proof

We will prove this theorem by induction. (When we construct the natural numbers in the next chapter, induction will be available right from the start.)

Specifically, we use induction on n. If $n = 0$, then the set $n = \{0, \ldots, n-1\}$ is the empty set, and the only set bijective with it is the empty set. So the induction starts.

Suppose that $k > 0$, and that the result is true with $n = k-1$. Now suppose that we are given a bijection f from $\{0, \ldots, k-1\}$ to $\{0, \ldots, m-1\}$. Suppose first that $f(k-1) = m-1$. Then the restriction of f to $\{0, \ldots, k-2\}$ is a bijection to $\{0, \ldots, m-2\}$. By the inductive hypothesis, $k-1 = m-1$, so $k = m$.

If $f(k-1) \neq m-1$, then we adjust f to produce a bijection which does satisfy this. Let $f(k-1) = a$ and $f(b) = m-1$, and define a function f' by the rule

$$f'(x) = \begin{cases} f(x) & \text{if } x \neq b, k-1; \\ a & \text{if } x = b; \\ m-1 & \text{if } x = k-1. \end{cases}$$

(See Figure 1.2.) Then f' is a bijection and $f'(k-1) = m-1$, and we conclude as before that $k = m$. Thus the result is proved for $n = k$, and the induction goes through.   □

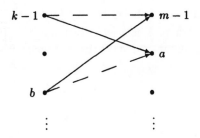

Fig. 1.2. Bijections

It follows that, for any set X, there is at most one natural number n such that X is bijective with n. We say that X is *finite* if such an n exists, and is *infinite* otherwise.

Sometimes the terms 'Peano finite' and 'Peano infinite' are used here. This is because an alternative definition was proposed by Dedekind: we say that a set X is *Dedekind infinite* if there is a bijection from X to a proper subset of itself, and is *Dedekind finite* if no such bijection exists.

Now a set which is Peano finite is Dedekind finite. For suppose that X is bijective with $\{0, \ldots, n-1\}$, and that X is bijective with a proper subset of itself. We may assume that $X = \{0, \ldots, n-1\}$. We show by induction that this is impossible: the induction clearly starts when $n = 0$. So suppose that $n > 0$, and that no bijection from $\{0, \ldots, n-1\}$ to a proper subset can exist, but there is a bijection f from $\{0, \ldots, n\}$ to a proper subset. Just as in the proof of the above theorem, we can modify f to ensure that $f(n-1) = n-1$. Then the restriction of f to $\{0, \ldots, n-2\}$ is a bijection from this set to a proper subset, contradicting the inductive hypothesis.

What about the converse? Suppose that X is Peano infinite. Then choose distinct elements $x_0, x_1, \ldots$ of X: we never get stuck since if $X \setminus \{x_0, \ldots, x_{n-1}\}$ were empty, then X would be bijective with $n = \{0, \ldots, n-1\}$, and so would be Peano finite. Then define a map $f : X \to X$ by the rule

$$f(x) = \begin{cases} x_{n+1} & \text{if } x = x_n; \\ x & \text{if } x \neq x_i \text{ for all } i. \end{cases}$$

Then f is a bijection from X to $X \setminus \{x_0\}$, and so X is Dedekind infinite.

This proof involves making infinitely many choices, and so it requires the Axiom of Choice. It is not true that Peano's and Dedekind's definitions of finiteness are equivalent without the assumption that this principle is true.

We will briefly touch on some counting results which hold in finite sets. For the remainder of this section, 'finite' means 'Peano finite'. We use the notation $|X| = n$ to mean that X is bijective with the set $n = \{0, \ldots, n-1\}$.

If n and m are natural numbers, we let $\binom{n}{m}$ denote the number of subsets of X of cardinality m, where X is a set of cardinality n. (Of course, it does not matter which set of cardinality n we choose.) As is standard,

$$\binom{n}{m} = \frac{n!}{m!(n-m)!}$$

where $n! = 1 \cdot 2 \cdots n$.

Theorem 1.12 (Principle of Inclusion and Exclusion)

Let $(A_i : i \in I)$ be a family of subsets of the finite set U, with $|I| = n$. For any non-empty $J \subseteq I$, let $A_J = \bigcap_{i \in J} A_i$, and let $A_\varnothing = U$. Then the number of elements of U which lie in none of the sets A_i is equal to

$$\sum_{J \subseteq I} (-1)^{|J|} |A_J|.$$

Proof

The summation in the theorem can be regarded as being made up of a contribution, for each element x of U, of the sum (over all J for which x lies in A_J) of $(-1)^{|J|}$. Now there are two cases:

- If x lies in none of the sets A_i, then the only set J such that $x \in A_J$ is $J = \varnothing$, and so the contribution of x to the sum is 1.

- Suppose that x does lie in some set A_i, and let $K = \{i \in I : x \in A_i\}$, with $k = |K|$. Then the sets J for which $x \in A_J$ are just the subsets of K; and the contribution of x to the sum is

$$\sum_{J \subseteq K} (-1)^{|J|} = \sum_{j=0}^{k} \binom{k}{j}(-1)^j = (1-1)^k = 0,$$

the second equality being an instance of the Binomial Theorem.

So the sum does indeed count the points lying in no set A_i.   □

Theorem 1.13

Let X and Y be finite sets with $|X| = m$ and $|Y| = n$. Then

(a) the number of functions from X to Y is n^m;

(b) the number of one-to-one functions from X to Y is $n(n-1)\cdots(n-m+1)$;

(c) the number of surjective functions from X to Y is

$$\sum_{j=0}^{n}(-1)^j \binom{n}{j}(n-j)^m.$$

Proof

We can assume that $X = \{0, \ldots, m-1\}$. A function f from X to Y is specified completely by the m-tuple

$$(f(0), \ldots, f(m-1))$$

of values. Now each value can be chosen to be any of the n elements of Y, and these choices are independent; so the number of functions is n^m, proving (a). If we require f to be one-to-one, then $f(0)$ may still be any of the n elements of X; $f(1)$ may be any element of X except $f(0)$ (giving $n-1$ choices); and so on, so that there are $n-m+1$ choices for $f(m)$; multiplying these numbers gives (b). (This argument is not valid if $m > n$, since we cannot have negative numbers of choices; but, if $m > n$, then there are no possible choices for $f(m-1)$, and so no one-to-one functions exist, and the formula in (b) correctly gives zero.)

Part (c) is proved using the Principle of Inclusion and Exclusion. We let U be the set of all functions from X to Y. For each element $y \in Y$, we let A_y be the set of all functions which do not take the value y. Now a function is onto if and only if it lies in none of the sets A_y, for $y \in Y$.

If $J \subseteq Y$, then A_J consists of all functions which do not take any value in the subset J. These are precisely the functions from X to $Y \setminus J$, and there are $(n - j)^m$ of them, if $j = |J|$. Now there are $\binom{n}{j}$ choices for the subset J of cardinality J, and each of them contributes a term $(-1)^j (n - j)^m$ to the sum in the Inclusion–Exclusion Principle. Applying this principle proves the result.   □

These results quantify, for finite sets, the facts that there is a one-to-one function from X to Y if and only if $|X| \leq |Y|$, and an onto function if and only if $|X| \geq |Y|$. However, it is not at all obvious that the formula in (c) yields zero when $m < n$. Try a few small values to check that it does.

1.7 Countable sets

We will briefly discuss countable sets in this section. The ideas here will be generalized later to arbitrary infinite sets.

Recall that a set is finite if it has the same cardinality as n, for some natural number n. We say that a set is *countable* if it has the same cardinality as the set $\mathbb{N}$.

Theorem 1.14

A set is at most countable if and only if it is finite or countable.

Proof

The statement involves a small adaptation of our earlier terminology: a set X is *at most countable* if there is an injective function from X to $\mathbb{N}$. Now if f is such a function, we modify it to form a function $g : X \to \mathbb{N}$ as follows: $g(x) = n$ if $f(x)$ is the nth element in the image of f. More formally, we define (by induction) $g(x) = 0$ if $f(x)$ is the least element of $f[X]$, and $g(x) = n$ if $f(x)$ is the smallest element of $f[X \setminus g^{-1}\{0, \ldots, n - 1\}]$. Now either this inductive procedure terminates because there are no elements left (in which case $f[X] = \{0, \ldots, n-1\}$, and X is finite), or it continues for ever and defines a bijection between X and $\mathbb{N}$.   □

We see that, if a set X is at most countable, then we can write it as $\{x_0, x_1, \ldots\}$, where either the list of elements stops at x_{n-1} for some n (if X has n elements), or it continues for ever (if X is countable).

The difficulties described at the end of Section 1.5 do not arise for sets which are at most countable. Any two such sets are comparable. (Two countable sets have the same cardinality; if X is finite and Y is countable then $|X| < |Y|$; and, if X has n elements and Y has m elements, then $|X| \leq |Y|$ if and only if $m \leq n$.) Moreover:

Theorem 1.15

There is a surjection from $\mathbb{N}$ to X if and only if X is at most countable.

Proof

We already know that the 'if' statement holds. So suppose that $g : \mathbb{N} \to X$ is a surjective function. For each $x \in X$, the set $g^{-1}(x) = \{n \in \mathbb{N} : g(n) = x\}$ is non-empty, so has a smallest element m_x. Now the function $f : X \to \mathbb{N}$ defined by $f(x) = m_x$ is injective since, if $x \neq y$, then $g(m_x) = x \neq y = g(m_y)$, so $m_x \neq m_y$. □

Various constructions applied to sets which are at most countable yield sets with the same properties.

Theorem 1.16

(a) The union of at most countably many at most countable sets is at most countable.

(b) The cartesian product of two at most countable sets is at most countable.

Proof

We observe first that a set is countable if and only if we can arrange its elements in an infinite sequence $(x_0, x_1, x_2, \ldots)$ so that each element occurs exactly once in the sequence. (This says no more or less than that the function $f : \mathbb{N} \to X$ defined by $f(n) = x_n$ is a bijection.)

We prove first that the cartesian product of two countable sets is countable. For this, it is enough to prove that $\mathbb{N} \times \mathbb{N}$ is countable, so we have to arrange the set of ordered pairs of natural numbers in a sequence. This we do as follows. First we break $\mathbb{N} \times \mathbb{N}$ into finite sets $S_0, S_1, S_2, \ldots$, where

$$S_k = \{(i, j) \in \mathbb{N} \times \mathbb{N} : i + j = k\}.$$

The set S_k contains just $k + 1$ pairs, namely

$$S_k = \{(0, k), (1, k - 1), \dots, (k, 0)\}.$$

Now we arrange the list by writing first the one element of S_0, then the two elements of S_1, then the three elements of S_2, and so on. (See Figure 1.3.)

$$
\begin{array}{ccccccccc}
(0,0) & & (0,1) & & (0,2) & & (0,3) & & (0,4) & \cdots \\
 & \swarrow & & \swarrow & & \swarrow & & \swarrow & \\
(1,0) & & (1,1) & & (1,2) & & (1,3) & & (1,4) & \cdots \\
 & \swarrow & & \swarrow & & \swarrow & & \swarrow & \\
(2,0) & & (2,1) & & (2,2) & & (2,3) & & (2,4) & \cdots \\
 & \swarrow & & \swarrow & & \swarrow & & \swarrow & \\
(3,0) & & (3,1) & & (3,2) & & (3,3) & & (3,4) & \cdots \\
 & \swarrow & & \swarrow & & \swarrow & & \swarrow & \\
(4,0) & & (4,1) & & (4,2) & & (4,3) & & (4,4) & \cdots \\
\vdots & & \vdots & & \vdots & & \vdots & & \vdots & \ddots
\end{array}
$$

Fig. 1.3. Arranging $\mathbb{N} \times \mathbb{N}$ in a sequence

It is possible to work out an explicit formula for the bijection f from $\mathbb{N} \times \mathbb{N}$ to $\mathbb{N}$, though (as we have stressed) this is not necessary. The element (i, j) is the $(i + 1)$st in S_{i+j}, and so is preceded by i elements of S_{i+j} together with all the elements of $S_0 \cup \dots \cup S_{i+j-1}$ (in number $1 + 2 + \dots + (i + j) = (i + j)(i + j + 1)/2$). So

$$f((i, j)) = (i + j)(i + j + 1)/2 + i.$$

Now assertion (b) of the theorem follows: for, if X and Y are at most countable, then there are countable sets X' and Y' containing X and Y; and $|X \times Y| \le |X' \times Y'|$.

Assertion (a) also follows easily. By a similar argument, it is enough to prove that the union of countably many countable sets is countable. So let $X_0, X_1, \dots$ be countable sets, and let $X_i = \{x_{i0}, x_{i1}, \dots\}$ for each i. Then the map f defined by $f((i, j)) = x_{ij}$ is a surjection from $\mathbb{N} \times \mathbb{N}$ to $\bigcup_{i \in \mathbb{N}} X_i$; so the union is countable. (This function need not be a bijection, since the sets X_i may not be disjoint.) $\qquad\square$

As a illustration of this result, we show that the set of finite subsets of $\mathbb{N}$ is countable. (By contrast, we know from Cantor's Theorem that the set of all subsets of $\mathbb{N}$ is not countable.) Let $\mathcal{P}_{\text{fin}} \mathbb{N}$ denote the set of finite subsets of $\mathbb{N}$. Then

$$\mathcal{P}_{\text{fin}} \mathbb{N} = \bigcup_{n \in \mathbb{N}} \mathcal{P}_n \mathbb{N},$$

where $\mathcal{P}_n \mathbb{N}$ is the set of n-element subsets of $\mathbb{N}$. Now $\mathcal{P}_0 \mathbb{N} = \{\varnothing\}$ has just one element. For $n > 0$, there is an injective map from $\mathcal{P}_n \mathbb{N}$ to $\mathbb{N}^n$: take any n-element set, and map it to the n-tuple obtained by writing its elements in increasing order. So $\mathcal{P}_n \mathbb{N}$ is countable, and $\mathcal{P}_{\text{fin}} \mathbb{N}$ is a countable union of at most countable sets, whence countable.

Here is another illustration. In the next section, we will outline the formal construction of the integers, rational, real and complex numbers; but we know enough about them to decide whether there are countably or uncountably many of each sort of number.

Theorem 1.17

The sets $\mathbb{Z}$ of integers and $\mathbb{Q}$ of rational numbers are both countable.

Proof

We have
$$\mathbb{Z} = \mathbb{N} \cup \{-n : n \in \mathbb{N}\},$$
so $\mathbb{Z}$ is the union of two countable sets, hence countable.

Any rational number can be written in its lowest terms as a/b, where a and b are coprime integers and $b > 0$; and this representation is unique. So the map $g : \mathbb{Q} \to \mathbb{Z} \times \mathbb{Z}$ given by $g(a/b) = (a, b)$ where a/b are as above, is an injection; and so $\mathbb{Q}$ is countable. $\square$

By contrast, we have:

Theorem 1.18

The set $\mathbb{R}$ of real numbers is not countable.

Proof

Suppose for a contradiction that $\mathbb{R}$ is countable. Then the subset $(0, 1)$ of $\mathbb{R}$ (the open unit interval) is also countable. So we can write its elements as a sequence:
$$(0, 1) = \{r_0, r_1, r_2, \ldots\}.$$
Each real number in the unit interval can be written as an infinite decimal (possibly terminating, in which case we let it continue forever with zeros). Let

$$r_i = 0.x_{i0}x_{i1}x_{i2}\ldots$$

be the decimal expansion of r_i, where $x_{ij} \in \{0, 1, 2, \ldots, 9\}$. Define

$$y_i = \begin{cases} 7 & \text{if } x_{ii} \neq 7, \\ 3 & \text{if } x_{ii} = 7. \end{cases}$$

Then $y_i \neq x_{ii}$ for all i. Let r be the real number given by

$$r = 0.y_0 y_1 y_2 \ldots$$

Then $r \neq r_i$, since their decimal expansions differ in the ith place. So r is a real number in the unit interval not included in the sequence, contrary to assumption.   □

1.8 The number systems

Leopold Kronecker said, 'God made the natural numbers: the rest is the work of man.' In the next chapter, we will see how we can construct the natural numbers themselves out of nothing, in mathematics built on set theory. In this section, we consider briefly how we build the rest of mathematics from the natural numbers. Later, we will look at Peano's alternative approach to the natural numbers, as an axiomatic system.

We assume, then, that we have the set $\mathbb{N}$ of natural numbers, on which the operations of addition and multiplication and the usual order relation are defined. Among the properties which hold are the following:

- addition and multiplication are commutative and associative;

- zero is the identity for addition, and 1 is the identity for multiplication;

- the *cancellation laws* hold for addition and for multiplication (except by zero): that is,

$$a + c = b + c \quad \text{implies} \quad a = b,$$
$$ac = bc, \ c \neq 0 \quad \text{implies} \quad a = b;$$

- if $a < b$, then $a + c < b + c$ and (if $c \neq 0$) $ac < bc$.

Subtraction is not everywhere defined; that is, the equation $a + x = b$ has a solution (for given a, b) only if $a \leq b$. Our first task is to extend the natural numbers to the *integers*, so that subtraction is everywhere defined. Accordingly, we have to add solutions to all equations of this form. Each ordered pair (a, b) should determine a unique integer x satisfying $a + x = b$. Accordingly, we will *represent* this x by the ordered pair. Different ordered pairs should determine the same x; so in fact x should be represented by an equivalence class of ordered pairs. What is the corresponding equivalence relation? In other words, when will $b - a = d - c$ hold? This equation can be written as $a + d = b + c$, which makes sense in $\mathbb{N}$. Accordingly we proceed as follows.

Define a relation $\sim$ on $\mathbb{N} \times \mathbb{N}$ by the rule that $(a, b) \sim (c, d)$ if and only if $a + d = b + c$. It is a simple exercise to check (using the rules given above) that $\sim$ is an equivalence relation. Now we define the *integers* to be the set

$$\mathbb{Z} = (\mathbb{N} \times \mathbb{N})/\sim$$

of equivalence classes of this relation. We let $[a, b]$ denote the equivalence class containing (a, b). Now we define addition, multiplication and order on $\mathbb{Z}$ as follows:

- $[a, b] + [c, d] = [a + c, b + d]$;
- $[a, b] \cdot [c, d] = [ac + bd, ad + bc]$;
- $[a, b] \leq [c, d]$ if and only if $a + d \leq b + c$.

Where do these definitions come from? They are obtained from our intended interpretation, that $[a, b]$ will be the integer $a - b$. For example, $(a - b)(c - d) = (ac + bd) - (ad + bc)$.

We have first to show that these definitions are good ones; that is, that a different choice of representatives of the equivalence classes would not change the object defined. For example, suppose that $(a, b) \sim (a', b')$ and $(c, d) \sim (c', d')$. Then a short calculation shows that $(a + c, b + d) \sim (a' + c', b' + d')$ and $(ac + bd, ad + bc) \sim (a'c' + b'd', a'd' + b'c')$.

Then it is not difficult to show that the usual arithmetic properties hold in $\mathbb{Z}$: for example, it is a commutative ring with identity and has no divisors of zero. (See Wallace [46], Chapters 3 and 5, for the concepts of ring theory.)

Moreover, the map that takes a to $[a, 0]$ is an injective function from $\mathbb{N}$ to $\mathbb{Z}$, and preserves addition, multiplication, and order. So, if we are working with $\mathbb{Z}$, we can now throw away our usual definition of $\mathbb{N}$ and regard a natural number as a special kind of integer (one which happens to satisfy $a \geq 0$).

The following constructions will be treated a bit more briefly; they are mostly similar to this one.

To obtain the rational numbers from the integers, we must add solutions to equations $ax = b$ with $a \neq 0$; and the equations $ax = b$ and $cx = d$ should have the same solution if $ad = bc$. So we define a relation $\sim$ on the set

$$\{(a, b) \in \mathbb{Z}^2 : a \neq 0\}$$

by the rule that $(a, b) \sim (c, d)$ if and only if $ad = bc$, and define the rational numbers to be the equivalence classes of this relation. Then we add, multiply and compare equivalence classes by the rules

- $[a,b] + [c,d] = [ad+bc, bd]$;

- $[a,b] \cdot [c,d] = [ac, bd]$;

- $[a,b] \leq [c,d]$ if and only if $abd^2 \leq b^2cd$.

(These are the rules for fractions!) Prove that the operations are well-defined and that the set $\mathbb{Q}$ is an ordered field. Also, the map from $\mathbb{Z}$ to $\mathbb{Q}$ taking a to $[a,1]$ is an injective function which preserves addition, multiplication and order, so we can regard $\mathbb{Z}$ as a subset of $\mathbb{Q}$.

This procedure is known to algebraists as constructing the *field of fractions* of an integral domain.

The ordered set of rational numbers has many 'gaps'; useful numbers like $\sqrt{2}$, π and e are missing. (We can approximate them as closely as we choose by rational numbers, but cannot express them exactly.) The construction of the real numbers is designed to 'fill these gaps'.

The construction is a little more complicated than earlier ones. As we have seen, it increases the cardinality, so it cannot be done just by taking equivalence classes of pairs. Two procedures are commonly used, involving *Cauchy sequences* and *Dedekind cuts* respectively. We outline the second method. The idea is that any real number r cuts the rationals into two sets, namely $\{x \in \mathbb{Q} : x \leq r\}$ and $\{x \in \mathbb{Q} : x > r\}$. Different real numbers give different cuts of $\mathbb{Q}$, since between any two real numbers there is a rational. So we can identify the real number with the corresponding cut.

A *Dedekind cut* is a partition of $\mathbb{Q}$ into two subsets L and R with the properties

- every element of L is smaller than every element of R;

- R has no least element.

Then $\mathbb{R}$ is the set of all Dedekind cuts. We write a cut as an ordered pair (L, R).

Defining the arithmetic is not quite straightforward. As a first attempt, we would put $(L,R) + (L',R') = (L+L', R+R')$, where

$$L + L' = \{x + y : x \in L, y \in L'\}.$$

However, this may not be a Dedekind cut, since we may be missing one rational. (If the corresponding reals are $\sqrt{2}$ and $-\sqrt{2}$, for example, then 0 would lie in neither set.) In this case, we add the missing rational to $L+L'$. Multiplication is even more complicated, since positive and negative numbers have to be handled separately. Order, fortunately, is easy: $(L,R) \leq (L',R')$ if and only if $L \subseteq L'$.

Now with some work it can be shown that $\mathbb{R}$ is an ordered field and satisfies the *Principle of the Supremum*:

Every non-empty set of real numbers which has an upper bound
has a least upper bound (or supremum).

From these facts, the standard theorems of real analysis can be derived.

The construction of $\mathbb{C}$ from $\mathbb{R}$ can be done most easily by taking a complex
number to be a pair of real numbers (namely, its real and imaginary parts).
That is, $\mathbb{C} = \mathbb{R} \times \mathbb{R}$. We define addition and multiplication by

- $(a, b) + (c, d) = (a + c, b + d)$;

- $(a, b) \cdot (c, d) = (ac - bd, ad + bc)$.

Now it can be shown that $\mathbb{C}$ is a field, and is *algebraically closed* (that is, any
polynomial equation over $\mathbb{C}$ has a solution).

1.9 Shoes and socks

The people up in Eppalock
Perambulate in one odd sock.
The other socks they hide away
To use as gifts on Christmas Day.

Michael Dugan and Walter Stackpool, *Nonsense Places: An Absurd
Australian Alphabet*

Bertrand Russell posed the following problem:

(a) Suppose that a drawer contains infinitely many pairs of shoes. Construct
a set containing one shoe from each pair.

(b) Same question for pairs of socks.

Part (a) of Russell's problem has an easy solution: you could simply take
the right shoe from each pair. Part (b), however, is more problematic, since
there is no natural rule for choosing one sock from each pair as there is for
shoes.

More formally, the question asks whether there is a set S containing one
element from each of an infinite set of two-element sets. This is rather different
from the problem about 'the set of all sets which are not members of themselves'
that we encountered earlier. It is not that S is 'too large' to be a set, since it
is a subset of something we know to be a set (the union of all the two-element
sets). Rather, it is that there is no obvious rule to guide the choice. There is
nothing self-contradictory in the assumption that the set exists, nor (somewhat
surprisingly) in the assumption that it doesn't exist.

Note that a set containing one sock from each pair is the image of a choice function for the family of pairs (an element of the cartesian product of the family). So Russell's problem, generalized, asks whether the cartesian product of any family of non-empty sets is non-empty.

We have already met the Axiom of Choice, which will guarantee that such sets do exist. However, this principle is not an *axiom* as Euclid used the term, a self-evident truth accepted by everybody. We will see that it has some paradoxical consequences! Because of its connection with cartesian products, Russell refers to it as the *Multiplicative Axiom*. We will examine it further in Chapter 6.

We could try to solve (b) in special cases. If the set of socks is countable, then we have an enumeration of it, as $\{s_0, s_1, \ldots\}$, and we could choose the sock in each pair with the smaller number. If the set of *pairs of socks* is countable, however, then we are in no better shape than we were without any information. This should make you look again at the proof of Theorem 1.16(a), that the union of countably many countable sets is countable. The proof actually required us to have an enumeration of each of the countable sets! So we were using the Axiom of Choice in this proof without realizing it, since having an enumeration of each pair of socks is exactly the same information as having a way of choosing one sock from each pair.

EXERCISES

1.1   Show that the empty set is a subset of every set.

1.2   Which of the following equations are true? If the equation is not true, is one side a subset of the other?

(a) $\bigcup \mathcal{P} X = X$.

(b) $\mathcal{P} \bigcup X = X$.

(c) $\bigcup \mathcal{P} X = \mathcal{P} \bigcup X$.

(d) $\mathcal{P}(X \times Y) = \mathcal{P} X \times \mathcal{P} Y$.

(e) $\mathcal{P}(X \cup Y) = \mathcal{P} X \cup \mathcal{P} Y$.

1.3   Prove that each of the following is *not* a suitable definition of the ordered pair (x, y):

(a) $(x, y) = \{x, \{y\}\}$.

(b) $(x, y) = \{\{x\}, \{y\}\}$.

1.4   Which of the following would be a suitable definition of the ordered triple (x, y, z)?

(a) $(x, y, z) = \{(x, y), (y, z)\}$.

(b) $(x, y, z) = ((x, y), (y, z))$.

(c) $(x, y, z) = \{\{x\}, \{x, y\}, \{x, y, z\}\}$.

1.5   This exercise describes a very fruitful source of equivalence relations in mathematics. (See Wallace [46], Chapter 6, for more details.) Let G be a group, and X a set. An *action* of G on X is a function $\mu : X \times G \to X$ satisfying the rules

- $\mu(x, gh) = \mu(\mu(x, g), h)$ for all $g, h \in G$, $x \in X$;

- $\mu(x, 1) = x$ for all $x \in X$, where 1 is the identity element of G.

(a) Prove that

- $\mu(\mu(x, g), g^{-1}) = \mu(\mu(x, g^{-1}), x) = x$ for all $x \in X$, $g \in G$.

(b) Define a relation $\sim$ on X by the rule that $x \sim y$ if and only if $\mu(x, g) = y$ for some $g \in G$. Show that $\sim$ is an equivalence relation. [*Note*: The equivalence classes of this relation are called the *orbits* of G in X (for the given action).]

(c) Show that each of the following equivalence relations on the set of all $m \times n$ real matrices arises from a group action:

- row-equivalence (A and B are row-equivalent if some sequence of elementary row operations transforms A to B);

- equivalence (A and B are equivalent if some sequence of elementary row and column operations transforms A to B);

- conjugacy, for $m = n$ (A and B are conjugate if $B = P^{-1}AP$ for some invertible matrix P);

- congruence, for $m = n$ (A and B are congruent if $B = P^{\mathsf{T}}AP$ for some invertible matrix P, where P^{T} is the transpose of P).

(d) If H is a subgroup of G, and H acts on G by *right multiplication* (that is, $\mu(x, h) = xh$), then the orbits of H are its *left cosets* in G.

(e) If G acts on itself by *conjugation* (that is, $\mu(x, g) = g^{-1}xg$), then the orbits of G are the *conjugacy classes*.

1.6   Suppose that μ is an action of the group G on the set X, as defined in the preceding exercise. Show that, for any $g \in G$, the function $x \mapsto \mu(x,g)$ is a bijection from X to X.

1.7   Show that the composition of injective functions is injective, and the composition of surjective functions is surjective.

1.8   Let $X \neq \varnothing$, let $f : X \to Y$ be a function, and let i_X and i_Y be the identity functions on X and Y respectively. Prove that

(a) f is injective if and only if there exists a function $g : Y \to X$ such that $f \circ g = i_X$;

(b) f is surjective if and only if there exists a function $h : Y \to X$ such that $h \circ f = i_Y$.

Where (if anywhere) have you used the Axiom of Choice in this proof?

1.9   Let $X \neq \varnothing$ and let $f : X \to Y$ be a function.

(a) Prove that f is injective if and only if $h_1 \circ f = h_2 \circ f$ implies $h_1 = h_2$, for any two functions $h_1, h_2 : Y \to X$.

(b) Prove that f is surjective if and only if $f \circ g_1 = f \circ g_2$ implies $g_1 = g_2$, for any two functions $h_1, h_2 : Y \to X$.

Where (if anywhere) have you used the Axiom of Choice in this proof?

1.10   Let R be a relation between X and Y. Define the *converse* of R to be the relation between Y and X defined by reversing all the pairs in R:

$$R^* = \{(y,x) : (y,x) \in R\}.$$

Show that the converse of a function f is a function if and only if f is bijective (in which case f^* is the inverse of f).

1.11   Let X and Y be finite sets, with m and n elements respectively. How many elements are there in each of the following sets?

(a) $\mathcal{P}X$.

(b) $X \times Y$.

(c) The set of relations from X to Y.

(d) The set $\prod_{y \in Y} X_y$, where $X_y = X$ for all $y \in Y$.

1.12   Show that

(a) any finite partially ordered set has a minimal element;

(b) any two (strict) total orders on a finite set are isomorphic;

(c) any (strict) partial order on a finite set X is contained in a (strict) total order on X.

1.13   Let R be a reflexive and transitive relation on a set X.

(a) Define a relation S on X by

$$S = \{(x, y) : (x, y), (y, x) \in R\}.$$

Show that S is an equivalence relation on X.

(b) Define a relation T on the set X/S of S-classes in X by

$$T = \{(S(x), S(y)) : (x, y) \in R\}.$$

Show that T is a non-strict order on X/S.

1.14   (a) Show that the cartesian product of finitely many copies of $\mathbb{N}$ is countable.

(b) Let X be a countable set. Show that the set X^* of all finite sequences of elements of X is countable.

(c) Prove that the set of *algebraic numbers* (those which satisfy some polynomial equation with integer coefficients) is countable. Prove that the set of *transcendental numbers* (those real numbers which are not algebraic) is uncountable.

1.15   This exercise completes our investigation of the cardinalities of the number systems.

(a) Show that there is a bijection between $\mathbb{R}$ and the open interval $(0, 1)$. [*Hint*: There is an analytic bijection.]

(b) Show that there is a bijection between the interval $(0, 1)$ and the interior of the unit square. [zemphHint: Interleave decimal expansions.]

(c) Deduce that $\mathbb{C}$ has the same cardinality as $\mathbb{R}$.

1.16   Let $(X, <)$ be a countable totally ordered set. Suppose that

(a) X is *dense*, that is, if $x < y$, then there exists z with $x < z < y$.

(b) X has no least or greatest element.

Prove that X is order-isomorphic to $\mathbb{Q}$.

   [*Hint*: Enumerate $X = (x_0, x_1, \ldots)$ and $\mathbb{Q} = (q_0, q_1, \ldots)$.
Now define, inductively, a map $f : X \to \mathbb{Q}$ as follows:

(a) $f(x_0) = q_0$.

(b) Suppose that $f(x_0), \ldots, f(x_{n-1})$ have been defined. Then the
n points $x_0, \ldots, x_{n-1}$ divide X into $n + 1$ intervals (including
two semi-infinite intervals); x_n lies in one of these intervals,
say (x_i, x_j). Now the corresponding interval $(f(x_i), f(x_j))$ in
$\mathbb{Q}$ is non-empty. Choose the rational number q_h with smallest
index in this interval, and define $f(x_n) = q_h$.

Prove that f is an order-preserving bijection.]

1.17   Use the same method to prove that any countable totally
ordered set is isomorphic to a subset of $\mathbb{Q}$.

1.18   For $n > 0$, define a function $f : \mathcal{P}_n(\mathbb{N}) \to \mathbb{N}$ by the rule

$$f(\{x_0, x_1, \ldots, x_{n-1}\}) = \binom{x_0}{1} + \binom{x_1}{2} + \cdots + \binom{x_{n-1}}{n},$$

where $x_0 < x_1 < \cdots < x_{n-1}$. Prove that f is a bijection.

1.19   Prove that the following two statements are equivalent.
(You may reason informally: when we discuss axioms for set theory
in Chapter 6, you can check whether your solution is valid in
axiomatic set theory. Note that statement (a) is what we have
defined as the Axiom of Choice; because of the equivalence, we
could take (b) instead if we preferred.)

(a) The cartesian product of any family of non-empty sets is
non-empty.

(b) Let P be a partition of X. Then there is a subset Y of X
which contains exactly one element from each member of P.

1.20   Use the Axiom of Choice to show that, if there is a surjection from Y to X, then there is an injection from X to Y.

2
Ordinal numbers

We have learned to pass with such facility from cardinal to ordinal number that the two aspects appear to us as one. To determine the plurality of a collection, that is, its cardinal number, we do not bother anymore to find a model collection with which we can match it – we *count* it ... The operations of arithmetic are based on the tacit assumption that *we can always pass from any number to its successor*, and this is the essence of the ordinal concept.

And so matching by itself is incapable of creating an art of reckoning. Without our ability to arrange things in ordered succession little progress could have been made. Correspondence and succession ... are woven into the very fabric of our number system.

Tobias Dantzig, *Number: The Language of Science* [12]

We learn numbers and counting as a process of succession. 'Eleven' has little real meaning to us except as 'the number after ten'. In this chapter, we use this process of succession to define the natural numbers – to do God's work, in Kronecker's phrase – starting from nothing (more precisely, the empty set) and progressing from one number to the next. As succession is the defining characteristic of natural numbers, so induction is the key proof technique. We can use it to define the arithmetic operations and to prove their basic properties.

But we do not have to stop there. By adding the principle of gathering up all the numbers so far constructed, we extend the ordinal number system into the infinite. We have transfinite induction to replace ordinary induction in our proofs. And, just as any finite set can be counted by a natural number, so any

well-ordered set can be counted by a unique ordinal number. Moreover, when
we come to define cardinal numbers in Chapter 6, we will see Dantzig's claim
borne out: the only sets for which a satisfactory theory of cardinal number has
been developed are those which can be well-ordered.

The ordinal numbers have another important role to play in the foundations.
The strategy for developing a consistent set theory avoiding Russell's Paradox
is to generate the sets out of nothing (that is, out of the empty set) in stages.
In this way, the complete totality of sets is never formed, and Russell's 'set of
all sets which are not members of themselves' is not defined. The stages in this
process cannot just proceed through the natural numbers but must continue
into the transfinite, at each stage admitting all subsets of the sets constructed
at previous stages. The ordinals give a precise description of these stages. This
is the most technical part of set theory which has to be developed before we
begin the axiomatic approach. Having secured our theory from contradiction
in this way, we follow standard mathematical practice by writing down axioms
which capture the theory we have developed.

2.1 Well-order and induction

A *well-order* on a set X is a total order $<$ on X having the property that every
non-empty subset of X has a least element (with respect to the restriction
of $<$). This grammatically monstrous term is a back-formation from the term
well-ordered set, which we apply to the ordered set $(X, <)$: strictly speaking we
should talk about a 'good order', but the term 'well-order' has become firmly
established and we continue to use it.

For example, the natural numbers (with the usual ordering) form a well-
ordered set: every non-empty subset of natural numbers has a least element.
Indeed, $(\mathbb{N}, <)$ is the simplest infinite well-ordered set. Any finite totally ordered
set is well-ordered.

You should recognize that the well-ordering of the natural numbers is closely
related to the idea of 'proof by induction'. One version of proof by induction
which is commonly used is the 'minimal counterexample' technique, where we
suppose that the proposition we are proving for all natural numbers is false,
and argue on the smallest number which is a counterexample, using the fact
that the proposition is true for all smaller numbers. Clearly this technique will
work in any well-ordered set.

Theorem 2.1

Let $(X, <)$ be a well-ordered set. Suppose that Y is a subset of X with the property that, for all $x \in X$, if it holds that $y \in Y$ for all $y < x$, then it holds that $x \in Y$. Then $Y = X$.

Proof

Suppose that $Y \neq X$, so that $X \setminus Y \neq \emptyset$. Since X is well-ordered, $X \setminus Y$ has a least element, say x. By definition, any $y < x$ does not lie in $X \setminus Y$, and so lies in Y. The hypothesis of the theorem now shows that $x \in Y$, contrary to our choice of x. So it cannot be that $Y \neq X$.   □

With a slight change, this becomes the *Principle of Induction*:

Theorem 2.2 (Principle of Induction)

Let $(X, <)$ be a well-ordered set. Let P be a property which may hold for elements of X. Suppose that, for all $x \in X$, if every element $y < x$ has property P, then x has property P. Then we conclude that every element of X has property P.

This follows on choosing Y to be the set of elements of X which have property P. Note that we don't have to do the base case of this induction: our hypotheses guarantee that the induction starts. For, if x is the smallest element of X, then there are no elements $y < x$, so vacuously all such elements have property P, whence x has property P by hypothesis.

2.2 The ordinals

We now develop the theory of *ordinals*, sometimes called *ordinal numbers*. These 'measure' well-ordered sets in the same way that natural numbers 'measure' finite sets. That is, given any finite set S, there is a unique natural number n such that a bijection exists between S and $\{1, 2, \ldots, n\}$: the number n is the *cardinality* of the set S. Inspired by this, we are going to prove the following theorem:

Theorem 2.3

Any well-ordered set is isomorphic to a unique ordinal.

The proof of this theorem is going to take the rest of this section. First, of course, we must define ordinals!

Given a totally ordered set $(X, <)$, and an element $a \in X$, we define the *section* X_a to consist of all elements of X which are less than a:

$$X_a = \{x \in X : x < a\}.$$

An *ordinal* is a well-ordered set $(X, <)$ with the property that $X_a = a$ for all $a \in X$. In other words, each element of X is the set of all its predecessors.

It is not immediately clear that ordinals exist. But we can see how they start. Let X be an ordinal. It has a least element a. Since a is the least element, we have $X_a = \varnothing$. But X is an ordinal, so $X_a = a$. Thus $a = \varnothing$. So the smallest element of any non-empty ordinal is $\varnothing$. Moreover, $\varnothing$ is (vacuously) an ordinal. Continuing, if $X \neq \{a\}$, then the subset $X \setminus \{a\}$ has a least element b; and we have $b = X_b = \{a\} = \{\varnothing\}$, so $\{\varnothing\}$ is the second ordinal. Continuing, we find that the next ordinal is $\{\varnothing, \{\varnothing\}\}$, and so on.

We will identify these 'starting ordinals' with the natural numbers – indeed, this will be our *definition* of natural numbers. We take

$$
\begin{aligned}
0 &= \varnothing \\
1 &= \{\varnothing\} = \{0\} \\
2 &= \{\varnothing, \{\varnothing\}\} = \{0, 1\} \\
3 &= \ldots = \{0, 1, 2\}
\end{aligned}
$$

and so on. In general, we have

$$n = \{0, 1, 2, \ldots, n-1\},$$

as we anticipated in the last chapter. So *every natural number is an ordinal.*

But the ordinals continue after the natural numbers leave off. If ω denotes the smallest ordinal which is not a natural number, then ω is the set of all natural numbers. Then the next ordinal after ω is $\omega \cup \{\omega\}$, and so on ...

Before we begin the proof of the theorem, it is convenient to have a result which shows that the above methods of constructing ordinals are typical.

Theorem 2.4

(a) If x is an ordinal, then so is $x \cup \{x\}$ (with $y < x$ for all $y \in x$).

(b) The union of a set of ordinals is an ordinal.

Proof

(a) The set $a = x \cup \{x\}$ (with order as specified in the statement of the theorem) has as sections all the sections of x and one additional one, namely a_x. But since all the elements of x are smaller than x, we have $a_x = x$. Moreover, for $y \in x$, we have $a_y = x_y = y$, since x is an ordinal.

(b) We defer the proof of this assertion until we know a little more about ordinals (after Lemma 2.7 below). $\qquad\qquad\square$

The proof of the theorem requires a series of technical lemmas.

Lemma 2.1

If $(X, <)$ is well-ordered, $Y \subseteq X$, and $f : X \to Y$ is an isomorphism, then $f(x) \geq x$ for all $x \in X$.

Proof

Induction. Let $E = \{x \in X : f(x) < x\}$. If $E \neq \varnothing$, then E has a least element, x_0 say. Then $f(x_0) < x_0$. Since f is an isomorphism, $f(f(x_0)) < f(x_0)$. But this shows that $f(x_0) \in E$, whereas $f(x_0)$ is smaller than the smallest element $x_0 \in E$, a contradiction. So $E = \varnothing$. $\qquad\qquad\square$

Lemma 2.2

There is at most one isomorphism between any two well-ordered sets.

Proof

Let $f, g : X \to Y$ be isomorphisms. Then $f \circ g^{-1}$ is an isomorphism from X to X, so $x \leq g^{-1}(f(x))$ for all $x \in X$ by Lemma 2.1, from which we see that $g(x) \leq f(x)$ since g is an isomorphism. But a similar argument shows that $f(x) \leq g(x)$ for all x, whence $f(x) = g(x)$ by antisymmetry. $\qquad\qquad\square$

Lemma 2.3

There is no isomorphism from a well-ordered set to a section of itself.

Proof

If $f : X \to X_a$ is an isomorphism, then $f(a) \in X_a$, so $f(a) < a$, contradicting Lemma 2.1.   □

Lemma 2.4

Let $(X, <)$ be a well-ordered set, and let $A = \{X_a : a \in X\}$ be the set of sections of X. Then $(A, \subset) \cong (X, <)$.

Proof

The isomorphism from X to A is given by $f(a) = X_a$. It is one-to-one since, if $a < b$, then $a \in X_b$ but $a \notin X_a$, so $X_a \neq X_b$; and clearly it is onto. Suppose that $a < b$. Then for any $x \in X_a$ we have $x < a$, so $x < b$, so $x \in X_b$; thus $X_a \subseteq X_b$. We already saw that these sets are not equal, so $X_a \subset X_b$. This shows that f is an isomorphism.   □

Lemma 2.5

Every section of an ordinal is an ordinal.

Proof

Let X be an ordinal and X_a a section of X. What is $(X_a)_b$ for $b \in X_a$? It consists of all the elements $x \in X_a$ which are less than b. But any x which is less than b is automatically less than a. So $(X_a)_b = X_b = b$, the last equality holding since X is an ordinal. We conclude that X_a is an ordinal.   □

Lemma 2.6

If X and Y are ordinals and $Y \subset X$, then X is a section of X.

Proof

Take a to be the least element of $X \setminus Y$. Then $X_a \subseteq Y$. Choose any $y \in Y$. If $a < y$, then $X_y = y = Y_y$ contains a, so $a \in Y$, contrary to assumption. Also $y = a$ is impossible since $y \in Y$ and $a \notin Y$. So $y < a$, and $y \in X_a$. So we have $Y \subseteq X_a$. We conclude that $Y = X_a$.   □

Lemma 2.7

Let X and Y be distinct ordinals. Then one is a section of the other.

Proof

First, $X \cap Y$ is an ordinal: for, if $a \in X \cap Y$, then $X_a = a = Y_a$, so all elements of a belong to both X and Y, and $a = (X \cap Y)_a$. Hence, by Lemma 2.6, $X \cap Y$ is a section of both X and Y.

Now suppose that $X \not\subset Y$ and $Y \not\subset X$. Then $X \cap Y = X_a$ for some $a \in X$, and $X \cap Y = Y_b$ for some $b \in Y$. But then

$$a = X_a = X \cap Y = Y_b = b \in X \cap Y,$$

a contradiction.   □

Proof of Theorem 2.4

First, note that any member of an ordinal is an ordinal. For, if x is an ordinal, $y \in x$, and $z \in y$, then $y = x_y$, so $y_z = (x_y)_z = x_z = z$.

Now let X be a set of ordinals. By the above remark, $A = \bigcup X$ is also a set of ordinals, and so there is an irreflexive and antisymmetric relation $<$ defined on A by the rule that $x < y$ if x is a section of y. Now $<$ is an order: for, if $x < y < z$, then $y = z_y$ and so $x = y_x = (z_y)_x = z_x$, so $x < z$. Now Lemma 2.7 shows that $<$ is a total order. Moreover, it is a well-order: for given a non-empty subset B of A, choose any $b \in B$; if it is not least, then all smaller elements are sections of b, and there is a least element among them since b is well-ordered. (This argument shows that any set of ordinals is well-ordered.) Finally, choose any $a \in A$; suppose that $a \in x \in X$. Then $a = x_a$, so all elements of a are in x, and hence in A, and we have $a = A_a$ as required. So A is an ordinal.   □

Lemma 2.8

If X and Y are isomorphic ordinals then $X = Y$.

Proof

Let $f : X \to Y$ be an isomorphism, and $E = \{x \in X : f(x) \neq x\}$. If $E = \varnothing$, then $X = Y$; so suppose not. If a is the least element of E, then $f(x) = x$ for all $x < a$, and so

$$a = X_a = Y_{f(a)} = f(a),$$

a contradiction.   □

Proof of Theorem 2.3

We *claim* the following:

> If $(X, <)$ is a well-ordered set such that, for each $a \in X$, the section X_a is isomorphic to an ordinal, then X is isomorphic to an ordinal.

Let us first see that this claim suffices to prove the theorem. Let $P(a)$ be the property 'X_a is isomorphic to an ordinal'. Then, assuming the claim, $P(a)$ holds for all $a \in X$, by induction; appealing to the claim one last time, X itself is isomorphic to an ordinal. Finally, if X is isomorphic to two ordinals, they are isomorphic to one another, and hence are equal, by Lemma 2.8. So the theorem is proved from the claim, and we now only have to establish the claim.   □

Proof of the Claim

Let $g_a : X_a \to Z(a)$ be an isomorphism for each $a \in X$, where $Z(a)$ is an ordinal. Note that $Z(a)$ and g_a are unique, by Lemmas 2.8 and 2.2. We can consider Z as a function on the set X. Let W be its range:

$$W = \{Z(a) : a \in X\}.$$

Now, if $x, y \in X$ and $x < y$, then $Z(x) \subset Z(y)$. For $Z(x)$ and $Z(y)$ are ordinals, and are not equal (since they are isomorphic to distinct sections of X); so one is a section of the other, by Lemma 2.7. It cannot be that $Z(y)$ is a section of $Z(x)$, else we could construct an isomorphism from X_y into its section X_x by composing g_y, the inclusion, and the inverse of g_x. So the function Z is a bijection, and indeed an isomorphism, from X to W (where W is ordered by inclusion). Thus W is well-ordered (being isomorphic to a well-ordered set). To finish the proof, we show that W is an ordinal. This holds because its members are ordinals, so any section W_a is equal to a.   □

The ordinals thus form a sequence of well-ordered sets, each contained in the next, which go on for ever. One variant of Russell's Paradox, known as the *Burali-Forti paradox*, is the following assertion:

Theorem 2.5

The ordinal numbers do not form a set.

Proof

If there were a set O consisting of the ordinal numbers, then it would itself be an ordinal number, and so it would be a member of itself; but it is obviously greater than each of its members!   □

Despite the fact that the ordinals do not form a set, it is still possible to think of them as if they were a set. In particular, the ordinals form an 'ordered class':

Theorem 2.6

For ordinals x and y, the following are equivalent:

(a) $x < y$;

(b) $x \in y$;

(c) $x \subset y$.

Moreover, exactly one of $x < y$, $x = y$, $y < x$ holds.

Proof

By Lemma 2.7, exactly one of $x \subset y$, $x = y$, $y \subset x$ holds. So the last assertion follows from the equivalence of (a) and (c). Now (a) requires a little interpretation: we write $x < y$ if x and y are members of some larger ordinal z for which this holds. Indeed, this doesn't depend on which ordinal z we take, since $x = z_x$ and $y = z_y$; and, if x is a section of y, then $z = y \cup \{y\}$ is an ordinal which contains both x and y. Now (a) and (b) are equivalent since $x < y$ if and only if $x \in z_y = y$. For (a) and (c), if $x < y$ then $x = z_x \subset z_y = y$; and if $x \not< y$ then $x = y$ or $y < x$, the latter implying $y \subset x$, so $x \not\subset y$.   □

It is also possible (and important) to do *induction* over all the ordinals:

Theorem 2.7

Let P be a property of ordinals. Suppose that, whenever x is an ordinal for which $P(y)$ holds for all ordinals $y < x$, then $P(x)$ holds. Then $P(x)$ holds for all ordinals.

Proof

To prove $P(x)$, it suffices to do induction over an ordinal containing x, such as $x \cup \{x\}$.   □

Although we have defined ordinals in a uniform way, it is convenient to subdivide them into three types; in many applications, the methods of handling these types are quite different.

The first type of ordinal consists only of zero, the smallest ordinal. (In inductive proofs, we are taught to make a separate argument for the base case where the value of the parameter is zero.)

The second type consists of *successor ordinals*. We observed above that, for any ordinal α, the set $\alpha \cup \{\alpha\}$ is also an ordinal, called the *successor* of α. The positive natural numbers are all successor ordinals: $n + 1$ is the successor of n. In general, the successor of α is the smallest ordinal which is greater than α. We write the successor of α as $s(\alpha)$

A non-zero ordinal λ is called a *limit ordinal* if it is the union of all its predecessors:

$$\lambda = \bigcup_{\alpha < \lambda} \alpha.$$

A successor ordinal is not a limit ordinal: for if $\lambda = \alpha \cup \{\alpha\}$, then the ordinals smaller than λ are all contained in α, and so is their union.

Theorem 2.8

Any non-zero ordinal is either a successor ordinal or a limit ordinal.

Proof

Suppose that λ is non-zero and is not a successor ordinal. Let

$$\mu = \bigcup_{\alpha < \lambda} \alpha.$$

Since a union of ordinals is an ordinal, we see that μ is an ordinal; and clearly $\mu \subseteq \lambda$, that is, $\mu \leq \lambda$. Suppose that $\mu < \lambda$. Then μ is one of the sets in the union defining itself; so it is the greatest ordinal less than λ. Since $\lambda > \mu$, we have $\lambda \geq s(\mu)$; and we cannot have strict inequality here, or else $s(\mu)$ would be in the family whose union is μ, that is, $s(\mu) \leq \mu$, which is clearly false. So $\lambda = s(\mu)$ is a successor ordinal, a contradiction. We conclude that $\lambda = \mu$ and so λ is a limit ordinal. $\qquad\square$

As we remarked earlier, many arguments about ordinals (especially induction arguments) require different methods for the cases of zero, successors, and limit ordinals. For example, we can re-formulate induction so that it looks more like ordinary induction (at least in the case of successor ordinals):

Theorem 2.9

Let P be a property of ordinals. Assume that

- $P(0)$ is true;

- $P(\alpha)$ implies $P(s(\alpha))$ for any ordinal α;

- if λ is a limit ordinal and $P(\beta)$ holds for all $\beta < \lambda$, then $P(\lambda)$ holds.

Then $P(\alpha)$ is true for all ordinals α.

Proof

We let Q be the property which holds at α if and only if $P(\beta)$ holds for all $\beta \le \alpha$. Now we verify the hypotheses of Theorem 2.7 for Q. Suppose that $Q(\beta)$ is true for all $\beta < \alpha$. Then *a fortiori*, $P(\beta)$ holds for all $\beta < \alpha$. If $\alpha = 0$ or α is a limit ordinal, it follows from the hypotheses of the theorem that $P(\alpha)$ holds. Suppose that $\alpha = s(\beta)$. Then $P(\beta)$ holds so, by hypothesis, $P(\alpha)$ holds.

Thus $P(\alpha)$ is true in all cases. Since we know that $P(\beta)$ is true for all $\beta < \alpha$, we now deduce that $Q(\alpha)$ holds.

By Theorem 2.7, $Q(\alpha)$ (and hence $P(\alpha)$) is true for all ordinals α. $\square$

2.3 The hierarchy of sets

Zermelo's hierarchy was an approach to set theory aimed at avoiding the paradoxes. With modifications suggested by Fraenkel, it is the approach most commonly used today. Zermelo's idea was to build the sets in well-ordered stages. If sets which are not members of themselves continue to appear at every stage, then there will be no stage at which they all exist and can be gathered into a set, so Russell's Paradox will not arise. The stages of the construction will be indexed by the ordinals (since any well-ordered set is order-isomorphic to a unique ordinal). Let V_α be the set of all sets constructed at stage α. Then the inductive definition is as follows:

$$\begin{aligned}
V_0 &= \varnothing \\
V_{s(\alpha)} &= \mathcal{P} V_\alpha \\
V_\lambda &= \bigcup_{\alpha < \lambda} V_\alpha \text{ for limit ordinals } \lambda.
\end{aligned}$$

Thus,

$$\begin{aligned}
V_1 &= \{\varnothing\}; \\
V_2 &= \{\varnothing, \{\varnothing\}\}; \\
V_3 &= \{\varnothing, \{\varnothing\}, \{\{\varnothing\}\}, \{\varnothing, \{\varnothing\}\}\};
\end{aligned}$$

V_4 is a set with sixteen elements; and so on.

The main content of Zermelo's approach is that this procedure gives us all sets. That is, every set is contained in V_α for some ordinal α. Symbolically, we can write

$$V = \bigcup_{\alpha \in \mathrm{On}} V_\alpha,$$

where V is the 'class' of all sets and On the 'class' of all ordinal numbers. (This is only an *aide-mémoire*, not a mathematical expression!)

We derive now a couple of facts about the Zermelo hierarchy, using transfinite induction. First, it really is a hierarchy: the sets get larger as we progress. (This is not obvious; each set is the power set of its predecessor, and most sets X don't satisfy $X \subseteq \mathcal{P}\,X$.)

Theorem 2.10

$V_\alpha \subseteq V_\beta$ for $\alpha < \beta$.

Proof

The proof actually requires a double induction, which we separate into two steps.

Step 1

We *claim* that it suffices to prove that $V_\alpha \subseteq V_{s(\alpha)}$ for all ordinals α. For suppose that this holds, and suppose that $\alpha < \beta$ and $V_\alpha \not\subseteq V_\beta$. Let β be the smallest such ordinal (for given α). Clearly $\beta \neq 0$, so there are two cases:

Case 1: β is a successor ordinal, say $\beta = s(\gamma)$. Then $\alpha \leq \gamma$, so

$$V_\alpha \subseteq V_\gamma \subseteq V_{s(\gamma)} = V_\beta,$$

the second inclusion following from the claim.

Case 2: β is a limit ordinal. Then $V_\alpha \subseteq V_\lambda$ for all $\alpha \leq \lambda < \beta$, so

$$V_\alpha \subseteq \bigcup_{\lambda < \beta} V_\lambda = V_\beta.$$

Step 2

We prove that $V_\alpha \subseteq V_{s(\alpha)}$ for all ordinals α by induction.

Case 1: $\alpha = 0$. Then V_α is the empty set, which is a subset of any set.

Case 2: $\alpha = s(\gamma)$ for some γ. Take $x \in V_\alpha$. Then

$$x \subseteq V_\gamma \subseteq V_{s(\gamma)} = V_\alpha,$$

so $x \in \mathcal{P} V_\alpha = V_{s(\alpha)}$.

Case 3: α is a limit ordinal. Take $x \in V_\alpha$. Then $x \in V_\delta$ for some $\delta < \alpha$; so

$$x \in V_{s(\delta)} \subseteq V_{s(\alpha)},$$

since $V_{s(\delta)} = \mathcal{P} V_\delta \subseteq \mathcal{P} V_\alpha = V_{s(\alpha)}$. $\qquad\square$

Theorem 2.11

For any ordinal α, we have $\alpha \subseteq V_\alpha$, and hence $\alpha \in V_{s(\alpha)}$.

Proof

Again the proof is by induction.

Case 1: $\alpha = 0 = \varnothing$: then α is a subset of any set!

Case 2: $\alpha = s(\gamma) = \gamma \cup \{\gamma\}$. Now $\gamma \subseteq V_\gamma \subseteq V_\alpha$, and $\gamma \in V_{s(\gamma)} = V_\alpha$, both by the induction hypothesis; so $\alpha \subseteq V_\alpha$.

Case 3: α is a limit ordinal. Then

$$\alpha = \bigcup_{\delta < \alpha} \delta \subseteq \bigcup_{\delta < \alpha} V_\delta = V_\alpha.$$

$\qquad\square$

There are two drawbacks with this simple approach to rigorous set theory. First, we appear to be defining sets in terms of ordinals, which are themselves sets: is our procedure not circular? Second, it is not easy, from this approach, to prove things about sets.

In Chapter 6, we will deduce from Zermelo's hierarchy a number of assertions about sets. We will then take these assertions as axioms for a formal theory of sets. As always in mathematics, axiomatization represents a development in a field which has already achieved some mathematical maturity. Zermelo's insight gives us confidence in our formal manipulations with the axioms.

2.4 Ordinal arithmetic

'Can you do Addition?' the White Queen asked. 'What's one and one and one and one and one and one and one and one and one and one?'

'I don't know,' said Alice. 'I lost count.'

'She can't do Addition,' the Red Queen interrupted.

Lewis Carroll, *Through the Looking-Glass, and what Alice found there*

The ordinals we have defined are a kind of numbers – indeed, they include the natural numbers – although they are designed for 'counting' well-ordered sets, not arbitrary sets. Accordingly, we would like to do arithmetic with them; in particular, to add and multiply them. There are two approaches to this.

First is the structural approach: we figure out how to 'add' and 'multiply' well-ordered sets, and use these to define the operations on ordinals. Essentially, we take the 'ordered sum' of two ordered sets to be their disjoint union, with each element of the first set preceding each element of the second set. Since the sets may not be disjoint (as will indeed happen if they are non-zero ordinals), we 'tag' them to make disjoint copies, as in the first proof of the Schröder–Bernstein theorem. (We take the tags to be the first two ordinals, $0 = \varnothing$ and $1 = \{\varnothing\}$, though in fact any two distinct tags would do.) For multiplication, we take the 'lexicographic product' (the cartesian product with the lexicographic order).

Definition 2.1

Let $(X, <_X)$ and $(Y, <_Y)$ be ordered sets. We define the *ordered sum* of these sets to be $(Z, <_Z)$, where

- $Z = (X \times \{0\}) \cup (Y \times \{1\})$;

- $(x_1, 0) <_Z (x_2, 0)$ if and only if $x_1 <_X x_2$;

- $(y_1, 1) <_Z (y_2, 1)$ if and only if $y_1 <_Y y_2$;

- $(x, 0) <_Z (y, 1)$ for all $x \in X$, $y \in Y$.

We define the *lexicographic product* of the sets to be $(W, <_W)$, where

- $W = X \times Y$;

- $(x_1, y_1) <_W (x_2, y_2)$ if $y_1 <_Y y_2$;

- $(x_1, y) <_W (x_2, y)$ if $x_1 <_X x_2$.

It can be shown that the ordered sum and lexicographic product of totally ordered sets are totally ordered sets; and ordered sum and lexicographic product of well-ordered sets are well-ordered sets.

Definition 2.2

Let α and β be ordinals. We define $\alpha + \beta$ to be the unique ordinal isomorphic to the ordered sum of α and β, and $\alpha \cdot \beta$ to be the unique ordinal isomorphic to their lexicographic product.

The second definition is more formal, using transfinite induction.

Definition 2.3

- $\alpha + 0 = \alpha$.

- $\alpha + s(\beta) = s(\alpha + \beta)$.

- If λ is a limit ordinal then $\alpha + \lambda = \bigcup_{\beta < \lambda} \alpha + \beta$.

- $\alpha \cdot 0 = 0$.

- $\alpha \cdot s(\beta) = \alpha \cdot \beta + \alpha$.

- If λ is a limit ordinal then $\alpha \cdot \lambda = \bigcup_{\beta < \lambda} \alpha \cdot \beta$.

It can be shown that this definition agrees with the previous one. (The proof, of course, is by induction.) The advantage of this approach is its flexibility. The last clause in the definition is essentially the same in both cases, and is easily modified for other situations. So if we want to define, for example, exponentiation of ordinals, we can proceed as follows:

Definition 2.4

- $\alpha^0 = 1$.

- $\alpha^{s(\beta)} = \alpha^\beta \cdot \alpha$.

- If λ is a limit ordinal then $\alpha^\lambda = \bigcup_{\beta < \lambda} \alpha^\beta$.

Now the *natural numbers* are just the ordinals less than ω, so we have incidentally defined them and shown how to add and multiply them. (The definitions are most easily obtained from Definition 2.3 by dropping the clauses about limit ordinals.) Now the properties of natural numbers that we used in Section 1.8 can be proved by induction: see Exercise 2.4. We use the usual notation for the natural numbers, so that, for example, $2 = \{\varnothing, \{\varnothing\}\}$. The first infinite ordinal (the set of all natural numbers) is usually denoted by ω.

Various properties of ordinal arithmetic can be proved (see Exercise 2.6 for some of these). Perhaps more interesting are the properties which are not true. For example,

$$1 + \omega = \omega \neq \omega + 1.$$

So the commutative law for addition and the cancellation law both fail. Indeed, if we take an infinite sequence and place a new element to the left, we still have an infinite sequence; but if we place a new element to the right, we get a different ordered set (one with a greatest element).

Ordinals soon grow to a point where it is not easy to imagine the resulting sets. For example, ω^2 is an infinite sequence of infinite sequences. Further along the sequence, we come to ω^3, ω^4, ..., and then

$$\omega^\omega = \bigcup_{n \in \mathbb{N}} \omega^n.$$

But we can continue, to reach ordinals like $\omega^{\omega^{\cdot^{\cdot^{\cdot^\omega}}}}$, (a 'tower' of $n+1$ omegas), which we denote by ω_n. More generally, we can define ω_α by

• $\omega_0 = \omega$,

• $\omega_{s(\alpha)} = \omega^{\omega_\alpha}$,

• $\omega_\lambda = \bigcup_{\alpha < \lambda} \omega_\alpha$ for limit ordinals λ.

Then ω_ω is an infinite 'tower' $\omega^{\omega^{\cdot^{\cdot^{\cdot}}}}$. Eventually, we reach the unimaginably large $\epsilon = \omega_{\omega_{\cdot_{\cdot_{\cdot}}}}$.

However, vast as these ordinals are, *they are all countable*, since each is a countable union of countable ordinals. Somewhere, even further down the line, lies the first uncountable ordinal ...

EXERCISES

2.1 Prove that the ordered sum and lexicographic product of totally ordered (resp., well-ordered) sets is totally ordered (resp., well-ordered).

2.2 Let X be any set, and define X^* to be the set of all finite sequences of elements of X. Prove that, if X can be well-ordered, then so can X^*. [*Hint*: $X^* = \bigcup_{n \in \mathbb{N}} X^n$; arrange the n-tuples in dictionary order.] Show that dictionary order on the set X^* is never a well-ordering if $|X| > 1$.

2.3   According to our definition, any natural number can be described in symbols as a sequence whose terms are the empty set $\varnothing$, opening and closing curly brackets { and }, and commas ,. For example, the number 4 is

$$\{\varnothing, \{\varnothing\}, \{\varnothing, \{\varnothing\}\}, \{\varnothing, \{\varnothing\}, \{\varnothing, \{\varnothing\}\}\}\}$$

with eight occurrences of $\varnothing$, eight of each sort of bracket, and seven commas. How many occurrences of each symbol are there in the expression for the number n?

2.4   Prove the properties of addition and multiplication of natural numbers used in Section 1.8.

2.5   Prove that the two definitions of ordinal addition and multiplication agree.

2.6   Prove the following properties of ordinal arithmetic:

(a) $(\alpha + \beta) + \gamma = \alpha + (\beta + \gamma)$.

(b) $(\alpha + \beta) \cdot \gamma = \alpha \cdot \gamma + \beta \cdot \gamma$.

(c) $\alpha^{\beta + \gamma} = \alpha^{\beta} \cdot \alpha^{\gamma}$.

2.7   (a) Show that, if $\gamma + \alpha = \gamma + \beta$, then $\alpha = \beta$.
    [*Hint*: The identity map from $\gamma + \alpha$ to $\gamma + \beta$ maps γ to γ and induces an isomorphism from α to β.]
    (b) Show that, if $\gamma \cdot \alpha = \gamma \cdot \beta$ and $\gamma \neq 0$, then $\alpha = \beta$.

2.8   Let $(X_i)_{i \in I}$ be a family of non-empty sets. Prove that, under either of the following conditions, the cartesian product $\prod_{i \in I} X_i$ is non-empty:

(a) $X_i = X$ for all $i \in I$;

(b) X_i is well-ordered for all $i \in I$.

2.9   Let X be a subset of the set of real numbers, which is well-ordered by the natural order on $\mathbb{R}$. Prove that X is finite or countable.
    [*Hint*: Let $X = \{x_\beta : \beta < \alpha\}$ for some ordinal α, and assume that $\beta < \gamma$ implies $x_\beta < x_\gamma$. Choose a real number q_β in the interval $(x_\beta, x_{s(\beta)})$ for all $\beta < \alpha$. Prove that these rational numbers are all distinct.]

2.10 (a) Show that any infinite ordinal can be written in the form $\lambda + n$, where λ is a limit ordinal and n a natural number.

(b) Show that any limit ordinal can be written in the form $\omega \cdot \alpha$ for some ordinal α.

2.11 Show that the set $\{m - \frac{1}{n} : m, n \in \mathbb{N}, m \geq 1, n \geq 2\}$ of rational numbers is isomorphic to ω^2. Find a set of rational numbers isomorphic to ω^3.

2.12 Show that there are uncountably many non-isomorphic countable ordinals. Using the fact that every countable totally ordered set is isomorphic to a subset of $\mathbb{Q}$ (see Exercise 1.16), give another proof of Cantor's Theorem that the power set of a countable set is uncountable.

3
Logic

'Contrariwise', continued Tweedledee, 'if it was so, it might be; and
if it were so, it would be; but as it isn't, it ain't. That's logic.'

Lewis Carroll, Alice's Adventures in Wonderland

Formal logic is another legacy of the set-theoretic paradoxes. In an attempt
to put mathematics on a sound footing, Hilbert proposed that the consistency
of a mathematical discipline could be established by translating the discipline
into a formal language which could be manipulated purely mechanically, by
arguments which would be constructive and would have no infinitary aspects.
Reasoning with these would show that no contradiction could be proved in the
original system. Though Hilbert's program was demolished by Gödel's famous
theorem (as we shall see in Chapter 5), formal logic has had a number of
important spin-offs within mathematics.

We must be clear at the start about one thing. The subjects of discussion in
formal logic are 'strings' or finite sequences of symbols from a fixed alphabet.
They have no meaning themselves. We speak of 'axioms', 'rules of inference',
'theorems' and 'proofs', but these words have specialized meanings which do
not entail their normal mathematical meanings. Only later do we set up inter-
pretations of the formulae in which the fact that a 'theorem' can be 'proved'
has any implications for the mathematical system we are really studying.

3.1 Formal logic

A *formal system*, then, involves the following:

- An *alphabet A*, a set of symbols.

- A set of *formulae*, each of which is a string of symbols from *A*. There must be a purely mechanical rule for deciding whether or not a given string is a formula. (The term 'well-formed formulae' is sometimes used.)

- A set of *axioms*, each axiom being a formula. The number of axioms may be finite or infinite, but there must be a purely mechanical rule for deciding whether or not a formula is an axiom.

- A set of *rules of inference*, each of which takes as 'input' a finite sequence of formulae and produces as output a formula. There must be a purely mechanical procedure for applying a rule of inference, so that, in particular, we can tell whether it has been correctly applied in a particular case.

A *proof* in a formal system is just a sequence of formulae such that each formula in the sequence either is an axiom or is obtained from earlier formulae in the sequence by applying a rule of inference. A *theorem* of the formal system is just the last formula in a proof.

It is possible to generate all the theorems of a formal system in a purely mechanical way. We could program a computer to start with the axioms and apply the rules of inference in all possible ways, printing out the results. Every theorem would eventually appear in the list. However, if a given formula had not appeared while we were watching the output of the computer, we would not know whether it was about to appear, or would appear in a million years, or would never appear. In general, we cannot expect that there will be a mechanical *decision procedure* which will decide whether a given formula is a theorem or not. This depends rather crucially on the formal system in question.

Having set up a formal system, we can treat it as a mathematical object in its own right, and reason mathematically about it. The conclusions we come to are often called *metatheorems*, as opposed to the theorems of the formal system; but from the mathematician's point of view they are theorems in the same sense as results in group theory or topology are theorems.

These ideas can be illustrated by an example which is so far removed from mathematics that the distinction between theorems and metatheorems is very clear. This example is due to Douglas R. Hofstadter, and is taken from his book, *Gödel, Escher, Bach: An Eternal Golden Braid* [22]. This is Hofstadter's *MU-system*.

The *alphabet* of the MU-system is the set {M, I, U}. A *formula* is any non-empty string of symbols. There is a single axiom, namely MI. There are four rules of inference:

Rule I: To any string ending with I, you may add U at the end.

Rule II: Given any string Mx beginning with M, you may duplicate the part after the M (obtaining Mxx).

Rule III: If three consecutive Is occur in a string, they may be replaced by U.

Rule IV: If two consecutive Us occur in a string, they may be deleted.

An example of a proof in the MU-system is

$$\text{MI, MII, MIIII, MUI, MUIU.}$$

We begin with the axiom MI, apply Rule II twice, then Rule III, and finally Rule I.

Note that the rules can be applied in many ways: at the third step above, we could have found MIU. Also, the same theorem can have many different proofs: we could prove MIU in one step by applying Rule I to the axiom MI.

Hofstadter asks: Is MU a theorem? If it is, then there must be a proof of it, and finding the proof settles the matter. But not finding the proof settles nothing; maybe we weren't clever enough!

In fact, MU is not a theorem. The argument that shows this is what we have described above as a metatheorem, and its proof illustrates a common technique for proving metatheorems: *by induction on the length of the proof.*

Theorem 3.1

In the MU-system, the number of occurrences of I in any theorem is not divisible by 3 (and in particular cannot be zero; so there is at least one I in any theorem).

Proof

As explained, we use induction on the length (the number of formulae) in the proof. The shortest proof has length 1, and consists just of the axiom MI, with one occurrence of I; since 1 is not divisible by 3, the base case of the induction is proved.

Now suppose that some theorem t has a proof of length n, and that the result is proved for all theorems with shorter proofs. Perhaps t is just the axiom; this is not forbidden, but if so, it has a proof of length 1, and we have considered this case already. So we may suppose that t follows from an earlier formula s in the proof by applying a rule. Let x be the number of occurrences of I in s. By the inductive hypothesis, x is not divisible by 3. Let y be the number of occurrences of I in t. We calculate y from x according to which rule was applied:

Rule I: $y = x$.

Rule II: $y = 2x$.

Rule III: $y = x - 3$.

Rule IV: $y = x$.

In each case, y is not divisible by 3. This completes the proof of the inductive step, and hence of the theorem. □

Hofstadter's choice of the form of this puzzle was not accidental; he was also teaching the distinction between a theorem and a metatheorem another way. It is a reference to one of the most famous *koans* of Zen Buddhism, *Joshu's Mu*:

A monk asked Joshu, a Chinese Zen master: 'Has a dog Buddha-nature or not?'
Joshu answered: '*Mu.*'

The answer is a negative term in Chinese, but it does not mean that Joshu has answered 'No.' Rather, his answer is neither 'Yes' nor 'No', but (roughly) 'The wrong question has been asked, or it has been asked from the wrong frame of mind.' Either 'Yes' or 'No' would be an answer in the system in which the questioner is operating; Joshu is commenting on that system from a position outside.

3.2 Propositional logic

Our goal in the next chapter is to give an account of 'first-order logic'. In the remainder of this chapter, we study 'propositional logic', sometimes called 'Boolean logic' or 'propositional calculus', in which many of the ideas occur in a simpler form.

We begin with a countable collection of *propositional variables*, $\{p_0, p_1, \ldots\}$. The language also contains symbols called *connectives*. The standard connectives are $\wedge$, $\vee$, $\neg$, $\rightarrow$ and $\leftrightarrow$ (read as 'and', 'or', 'not', 'implies' and 'if and only if' respectively). Also, there are left and right brackets (to make the formulae unambiguous). The specification of formulae is as follows.

• A propositional variable is a formula.

- If ϕ and ψ are formulae, then so are $(\neg\phi)$, $(\phi \vee \psi)$, $(\phi \wedge \psi)$, $(\phi \rightarrow \psi)$, and $(\phi \leftrightarrow \psi)$.

We require that any formula is given by the preceding rules. Note that it is possible to analyse a string of symbols to decide whether or not it is a formula in a purely mechanical way; if it is a formula, then it has a unique 'parsing', that is, there is a unique way in which it is built up by the given rules.

For example, we check that the following sequence is a formula:

$$(((\neg p_1) \rightarrow (p_1 \wedge p_2)) \vee (p_1 \wedge (p_2 \leftrightarrow p_3))).$$

If we strip off the outside brackets, we find that it has the shape $(\phi \vee \psi)$, where

$$\phi \text{ is } ((\neg p_1) \rightarrow (p_1 \wedge p_2)), \text{ and}$$
$$\psi \text{ is } (p_1 \wedge (p_2 \leftrightarrow p_3)).$$

Then ϕ has the shape $(\alpha \rightarrow \beta)$, where α is $(\neg p_1)$ and β is $(p_1 \wedge p_2)$. One further step reduces both α and β to propositional variables. The formula ψ is handled similarly.

The parsing of a well-formed formula can be represented by a tree. As in logic and computer science, trees grow upside down, with the root at the top, and the leaves at the bottom. Each leaf is labelled with a propositional variable p_i. Each node which is not a leaf has a 'type', which is one of the logical connectives. A node of type $\neg$ has one descendant; if the descendent has the label ϕ, then the node itself has label $(\neg\phi)$. Any other node has two descendants, and its label is the formula in which the labels of its descendants are combined by this connective. Figure 3.1 shows the tree, with the types and labellings, corresponding to the formula

$$(((\neg p_1) \rightarrow (p_1 \wedge p_2)) \vee (p_1 \wedge (p_2 \leftrightarrow p_3))).$$

Note how the structure of the tree reflects the process of parsing the formula.

Before we proceed to describe a deduction system for propositional logic, we explain what is known as the *semantics* of this logic. A formula of propositional logic is more than just a meaningless collection of symbols; it can represent a logical combination of facts about the universe. We interpret the propositional variables as basic statements. For example, if we are studying a particular group G, the variable p_0 might stand for the statement 'G is abelian', p_1 for the statement 'all elements of G have order 1 or 2', p_2 for 'G is finite', and so on. Then $(p_1 \rightarrow p_0)$ would be the statement 'if all elements of G have order 1 or 2, then G is abelian' (which happens to be true, regardless of whether p_0 and p_1 are true or false for the particular group we are considering), and $(\neg p_2)$ is the statement 'G is infinite'. As we see, if we know whether the basic

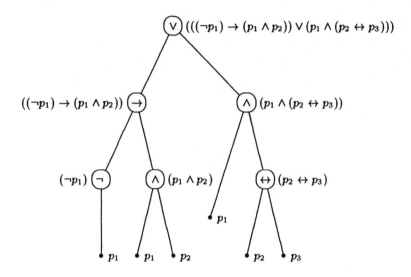

Fig. 3.1. A tree to parse a formula

statements p_i are true or false, we can in principle decide whether any such logical compound of them is true or false. The way in which this is done defines the semantics of propositional logic.

Any formula, which involves the propositional variables $p_0, \ldots, p_{n-1}$, can be used to define a function of n variables, that is, a function from the set $\{T, F\}^n$ to $\{T, F\}$. We can specify the semantics using the same scheme which is used to build the formula.

In order to make this precise, we define below a *valuation* to be a function v from the set of all formulae to the set $\{T, F\}$. The valuation maps each propositional variable p_i to a truth value, which we take to be the truth value of the corresponding basic proposition. We also have to specify how the valuation behaves as formulae are built up. This is done by truth tables as given in Table 3.1. For example, the truth table for $\wedge$ means that, if $v(\phi) = T$ and $v(\psi) = T$, then $v((\phi \wedge \psi)) = T$; otherwise, $v((\phi \wedge \psi)) = F$.

Applying a valuation to a formula can also be done using the tree associated with the formula. The valuation assigns truth values to the leaves of the tree; then we work upwards to the root, each node behaving as a function of one or two variables and feeding its output to the node above. In the example given earlier, if $v(p_1) = v(p_2) = T$ and $v(p_3) = F$, then v gives the value F, T, F to the three non-leaf nodes at the second level down from the root, T and F to the two nodes at the first level, and finally T to the root.

We say that ϕ is a *tautology* if $v(\phi) = T$ for all valuations v, and is a

ϕ	ψ	$(\phi \wedge \psi)$
T	T	T
T	F	F
F	T	F
F	F	F

ϕ	ψ	$(\phi \vee \psi)$
T	T	T
T	F	T
F	T	T
F	F	F

ϕ	$(\neg\phi)$
T	F
F	T

ϕ	ψ	$(\phi \rightarrow \psi)$
T	T	T
T	F	F
F	T	T
F	F	T

ϕ	ψ	$(\phi \leftrightarrow \psi)$
T	T	T
T	F	F
F	T	F
F	F	T

Table 3.1. Truth tables

contradiction if $v(\phi) = \mathsf{F}$ for all valuations v; and that ϕ is a *logical consequence* of a set Σ of formulae if every valuation v which satisfies $x(\sigma) = \mathsf{T}$ for all $\sigma \in \Sigma$, also satisfies $v(\phi) = \mathsf{T}$. In our group-theoretic example above, the formula $(p_2 \vee (\neg p_2))$ is a tautology; $(p_1 \rightarrow p_0)$ happens to be true for the particular p_0 and p_1 (even if we have no information about the group G), but in other choices of p_0 and p_1 it would be false, so it is not a tautology. We feel that the statement 'A group is either finite or infinite' is a logical truth, whereas the statement 'A group all of whose elements have order 1 or 2 is abelian' is merely a theorem of group theory.

Now we come to the definition of the formal deduction system for propositional logic. For this, we use only the connectives $\neg$ and $\rightarrow$, since all the others can be expressed in terms of these. Specifically, if we replace all occurrences of $(\phi \vee \psi)$ by $((\neg\phi) \rightarrow \psi)$, occurrences of $(\phi \wedge \psi)$ by $(\neg(\phi \rightarrow (\neg\psi)))$, and occurrences of $(\phi \leftrightarrow \psi)$ by $(\neg((\phi \rightarrow \psi) \rightarrow (\neg(\psi \rightarrow \phi))))$, then the value assigned to the formula by any valuation is not affected. This is shown by some easy calculations with truth tables, of which Table 3.2 is an example.

The formal system is specified as follows:

There are three 'schemes' of *axioms*, namely:

(A1) $(\phi \rightarrow (\psi \rightarrow \phi))$

(A2) $((\phi \rightarrow (\psi \rightarrow \theta)) \rightarrow ((\phi \rightarrow \psi) \rightarrow (\phi \rightarrow \theta)))$

(A3) $(((\neg\phi) \rightarrow (\neg\psi)) \rightarrow (\psi \rightarrow \phi))$

Each of these formulae is an axiom, for all choices of formulae ϕ, ψ, θ.

ϕ	ψ	$(\neg\psi)$	$(\phi \to (\neg\psi))$	$(\neg(\phi \to (\neg\psi)))$	$(\phi \wedge \psi)$
T	T	F	F	T	T
T	F	T	T	F	F
F	T	F	T	F	F
F	F	T	T	F	F

Table 3.2. Equivalence of $(\neg(\phi \to (\neg\psi)))$ and $(\phi \wedge \psi)$

There is only one *rule of inference*, namely *Modus Ponens*: From ϕ and $(\phi \to \psi)$, infer ψ.

The definition of a proof is exactly as usual in a formal system: that is, a sequence of formulae such that each is an instance of an axiom or follows from earlier formulae by a rule of inference. We define, more generally, a *proof of ϕ from Σ*, where Σ is a set of formulae and ϕ is a formula, to be a sequence of formulae in which each formula is an axiom *or a member of Σ* or follows from earlier formulae in the sequence by a rule of inference. (This is similar to forming a new formal system by adding Σ to the set of axioms, except that we don't require that Σ should be mechanically recognizable. Later, this freedom will be used to add to the purely logical axioms the specific axioms for some branch of mathematics such as group theory.)

When we write out a proof, we precede every formula by the symbol $\vdash$, denoting that it has been proved. If it is a proof from the set Σ, we write $\Sigma \vdash$ on the left.

As an example of the workings of the formal system, we prove one theorem and one important metatheorem.

Theorem 3.2

For any formula ϕ, the formula $(\phi \to \phi)$ is a theorem of propositional logic.

Proof

The formal proof is as follows:

$$\vdash \quad ((\phi \to ((\phi \to \phi) \to \phi)) \to ((\phi \to (\phi \to \phi)) \to (\phi \to \phi)))$$
$$\vdash \quad (\phi \to ((\phi \to \phi) \to \phi))$$
$$\vdash \quad ((\phi \to (\phi \to \phi)) \to (\phi \to \phi))$$
$$\vdash \quad (\phi \to (\phi \to \phi))$$
$$\vdash \quad (\phi \to \phi)$$

Here, the first formula is an instance of Axiom (A2), taking ϕ, ψ, θ to be ϕ,

$(\phi \to \phi)$ and ϕ respectively. The second is an instance of (A1), taking ϕ and ψ to be ϕ and $(\phi \to \phi)$. The third follows from the first two by *Modus Ponens*. The fourth is another instance of (A1), this time with ϕ, ψ and θ all set equal to ϕ. The fifth then follows from the third and fourth by *Modus Ponens*. □

You should note the extreme tediousness of even the most trivial deductions in this formal system. This is partly why we develop metatheorems, which can simplify the proofs.

Theorem 3.3 (Deduction Theorem)

Suppose that ϕ can be deduced from a set $\Sigma \cup \{\psi\}$ of formulae. Then $(\psi \to \phi)$ can be deduced from Σ.

Proof

It is convenient to note first that (A1) and *Modus Ponens* show immediately that, if we can deduce ϕ from any set of hypotheses, then we can deduce $(\psi \to \phi)$ from the same hypotheses.

Our argument will go by induction on the length of the proof (the number of formulae in the sequence). To do the base case, suppose that there is a one-line proof of ϕ from $\Sigma \cup \{\psi\}$, then ϕ is an axiom, a member of Σ or is ψ. In the first two cases, the hypothesis ψ is not used, and the remark at the start of our proof applies. If ϕ is the same as ψ, we are asked to prove $(\phi \to \phi)$, and this we did in the preceding theorem.

Now suppose that the proof of ϕ has more than one step, and assume the result for formulae with shorter proofs. If ϕ is an axiom or a member of $\Sigma \cup \{\psi\}$, then it could have been proved in one step, and we have already given the proof in this case.

So we may suppose that ϕ is deduced from θ and $(\theta \to \phi)$ by *Modus Ponens*, where θ and $(\theta \to \phi)$ occur earlier in the proof. Then θ and $(\theta \to \phi)$ have shorter proofs than ϕ. By the induction hypothesis, $(\psi \to \theta)$ and $(\psi \to (\theta \to \phi))$ can be proved from Σ. From these two formulae it is a simple matter (left to the reader) to prove $(\psi \to \phi)$ (using (A2) and *Modus Ponens* twice).

This completes the proof of the Deduction Theorem. □

Here is an example of use of the Deduction Theorem to simplify a proof. We prove that $((\neg\phi) \to (\phi \to \psi))$ is a theorem. It will suffice to deduce $(\phi \to \psi)$ from the set $\{(\neg\phi)\}$, and then apply the Deduction Theorem. Now from the given set of hypotheses, we have:

$$\{(\neg\phi)\} \quad \vdash \quad ((\neg\phi) \to ((\neg\psi) \to (\neg\phi)))$$

$$\{(\neg\phi)\} \quad \vdash \quad (\neg\phi)$$
$$\{(\neg\phi)\} \quad \vdash \quad ((\neg\psi) \to (\neg\phi))$$
$$\{(\neg\phi)\} \quad \vdash \quad (((\neg\psi) \to (\neg\phi)) \to (\phi \to \psi))$$
$$\{(\neg\phi)\} \quad \vdash \quad (\phi \to \psi)$$
$$\vdash \quad ((\neg\phi) \to (\phi \to \psi))$$

The first is an instance of (A1); the second, a hypothesis; the third follows by *Modus Ponens*; the fourth is an instance of (A3); the fifth follows; and the last invokes the Deduction Theorem, as explained.

3.3 Soundness and completeness

We now have two completely different ways of looking at propositional logic. On the one hand, it is a formal system, and we have notions of proof, theorem, and so on. On the other hand, the semantics of valuations gives some kind of meaning to the formulae; they are no longer just strings of symbols, but may be logically valid, or logical consequences of sets of formulae. Not surprisingly, it turns out that the notion of a theorem (in the formal system) coincides with that of a tautology or logically valid formula (one given the value T by every valuation): the axioms and rules of inference were chosen precisely so that this would be the case! This is the content of the *Propositional Soundness and Completeness Theorem*. Here, the term 'soundness' refers to the fact that theorems in the formal system are 'true' (that is, logically valid); 'completeness' refers to the fact that all true statements can be proved. In fact, the theorem is more general than we have just suggested. The notions of 'provable from Σ' and 'a logical consequence of Σ' also turn out to be equivalent, for any set Σ of formulae.

In the theorem below, we *assume* that we have a finite or countable collection of propositional variables p_i in our formal system. We say that a set Σ of formulae is *inconsistent* if there is a formula ψ such that both ψ and $(\neg\psi)$ can be deduced from Σ; if no such formula exists, then Σ is *consistent*.

Theorem 3.4 (Propositional Soundness and Completeness Theorem)

(a) A formula is a tautology if and only if it is a theorem (that is, provable).

(b) A formula is a logical consequence of a set Σ of formulae if and only if it is provable from Σ.

(c) Σ is consistent if and only if there is a valuation v satisfying $v(\sigma) = \mathsf{T}$ for all $\sigma \in \Sigma$.

Proof of Soundness

First we show that the axioms are tautologies and that *Modus Ponens* preserves truth (in the sense that if $v(\phi) = \mathsf{T}$ and $v((\phi \to \psi)) = \mathsf{T}$ then $v(\psi) = \mathsf{T}$). Both these assertions are simple exercises with truth tables.

Using this, it follows (by induction on the length of the proof) that if ϕ is a theorem then it is a tautology – this is true for one-line proofs (since axioms are tautologies), and anything deduced from tautologies using *Modus Ponens* is a tautology. More generally, if ϕ is provable from Σ, then any valuation which gives the value T to all formulae in Σ also gives the value T to ϕ. In particular, if there is a valuation v giving the value T to Σ, then Σ must be consistent (since no valuation can make both ψ and $(\neg\psi)$ true). This proves the reverse implications in the theorem (the Soundness Theorem). It remains to prove the forward implications (the Completeness Theorem), to which we now turn.

Proof of Completeness

Note first that (a) is a special case of (b), and (b) follows from (c) as follows. Suppose that (c) holds, and that ϕ is a logical consequence of Σ. Then $\Sigma \cup \{(\neg\phi)\}$ cannot be satisfied, so it is possible to prove both ψ and $(\neg\psi)$ from it, for some ψ. By the Deduction Theorem, it is possible to prove $((\neg\phi) \to \kappa)$ from Σ, where κ is either ψ or $(\neg\psi)$. From this, it is just a few steps to prove ϕ from Σ. This is most easily seen by noting that the formula

$$(((\neg\phi) \to \psi) \to (((\neg\phi) \to (\neg\psi)) \to \phi))$$

is a theorem (see Exercise 3.3).

Now (c) is proved as follows. We claim that, if Σ is consistent, then for any formula ϕ, either $\Sigma \cup \{\phi\}$ or $\Sigma \cup \{(\neg\phi)\}$ is consistent. For suppose not; then a contradiction κ can be deduced from both of these sets, so both $(\phi \to \kappa)$ and $((\neg\phi) \to \kappa)$ can be deduced from Σ. Since $((\phi \to \kappa) \to (((\neg\phi) \to \kappa) \to \kappa))$ is an instance of a theorem (Exercise 3.3), κ can be deduced from Σ, contrary to assumption.

Now the set of all formulae is countable. So Σ can be enlarged to a maximal consistent set Σ^+ as follows: Enumerate the formulae $\phi_0, \phi_1, \ldots$, and at the nth step add ϕ_n to Σ if the result is consistent, $(\neg\phi_n)$ otherwise. The resulting set Σ^+ has the property that, for any formula ϕ, either $\phi \in \Sigma^+$ or $(\neg\phi) \in \Sigma^+$. Clearly not both ϕ and $(\neg\phi)$ can belong to Σ^+, or it would not be consistent.

Define a valuation v by the rule that

$$v(p_i) = \begin{cases} \mathsf{T} & \text{if } p_i \in \Sigma^+, \\ \mathsf{F} & \text{otherwise.} \end{cases}$$

We claim that $v(\phi) = \mathsf{T}$ if and only if $\phi \in \Sigma^+$. The proof is by induction on the length of the formula ϕ: the induction begins because the claim is true for the formula p_i. If ϕ contains more than one symbol, then there are two cases:

- ϕ is $(\neg\psi)$. Now

$$v(\phi) = \mathsf{T} \Leftrightarrow v(\psi) = \mathsf{F} \Leftrightarrow \psi \notin \Sigma^+ \Leftrightarrow \phi \in \Sigma^+,$$

  where the second equivalence holds by the induction hypothesis.

- ϕ is $(\psi \to \theta)$. Now

$$\begin{aligned} v(\phi) = \mathsf{T} \quad &\Leftrightarrow \quad v(\psi) = \mathsf{F} \text{ or } v(\theta) = \mathsf{T} \\ &\Leftrightarrow \quad \psi \notin \Sigma^+ \text{ or } \theta \in \Sigma^+ \\ &\Leftrightarrow \quad (\neg\psi) \in \Sigma^+ \text{ or } \theta \in \Sigma^+. \end{aligned}$$

Suppose that $v(\phi) = \mathsf{T}$ and $\phi \notin \Sigma^+$. Then $(\neg\phi) \in \Sigma^+$. Since both

$$((\neg\psi) \to (\psi \to \theta))$$

and

$$(\theta \to (\psi \to \theta))$$

are theorems, both ϕ and $(\neg\phi)$ can be deduced from Σ^+, and Σ^+ is inconsistent. On the other hand, suppose that $v(\phi) = \mathsf{F}$ and $\phi \in \Sigma^+$. Then $\psi, (\neg\theta) \in \Sigma^+$. Now

$$(\psi \to ((\neg\theta) \to (\neg(\psi \to \theta))))$$

is a theorem (Exercise 3.3 again), and so both ϕ and $(\neg\phi)$ can be deduced from Σ^+, and again Σ^+ is inconsistent. We conclude that

$$v(\phi) = \mathsf{T} \Leftrightarrow (\phi \in \Sigma^+).$$

This completes the inductive step.

Now, since $\Sigma \subseteq \Sigma^+$, we have found a valuation satisfying Σ, as required. The proof is complete. $\square$

We can look at this result the other way around. Given a formula ϕ, is it a theorem of propositional logic? We have seen that proofs in this formal system are tedious to write and difficult to find, so our inability to prove ϕ doesn't allow us to conclude that it is not a theorem. The answer is that truth tables provide a mechanical *decision procedure* for the formal system, that is, a method for determining mechanically whether or not a given formula is a theorem. Write out the truth table for ϕ. If every value is T (that is, if ϕ is a tautology), then ϕ is a theorem; if not, it is not.

At this point, you may wonder what the use of the formal system is, since it seems to give us the same information as truth tables but less efficiently. The answer is rather unexpected. We will now prove the *Compactness Theorem*. We see in the proof how soundness and completeness are used in a crucial way. This result turns out to have important mathematical consequences.

We say 'Σ is *satisfiable*' as shorthand for 'there exists a valuation v such that $v(\sigma) = \mathsf{T}$ for all $\sigma \in \Sigma$'.

Theorem 3.5 (Propositional Compactness Theorem)

If every finite subset of Σ is satisfiable, then Σ is satisfiable.

For, by Theorem 3.4, Σ is satisfiable if and only if no contradiction can be derived from it; and a proof of a contradiction is a finite sequence of formulae, so can only use a finite number of formulae of Σ (and hence is a proof of a contradiction from some finite subset of Σ).

As an application, we give a proof promised in the Preface: the deduction of the infinite version of the Four-Colour Theorem from the finite.

A *plane map* is a plane figure consisting of a number of *countries*, each country a simply-connected region bounded by a simple closed curve, any two countries meeting (if at all) on part of their common boundary. A *colouring* of a map is a map from the set of countries of the map to a set of *colours*, having the property that if two countries have a common boundary (not just consisting of isolated points) then they are assigned different colours. If some of this talk about 'simply-connected regions bounded by simple closed curves' is a bit mysterious, see the discussion by Lakatos [32] emphasising the care which has to be taken in the formulation of such questions.

Kenneth Appel and Wolfgang Haken showed in 1976, with the help of about 2000 hours of computing on a mainframe machine, that 'four colours suffice':

Theorem 3.6 (Four-Colour Theorem for finite maps)

Any finite plane map can be coloured with four colours, so that adjacent countries are given different colours.

Theorem 3.7 (Four-Colour Theorem in general)

Any plane map can be coloured with four colours, so that adjacent countries are given different colours.

Proof

You might think that we could just keep extending a colouring of the finite portion of the map, but this is not valid; we might get stuck at some point. (It is exactly for this reason that the finite theorem was so difficult to prove.)

Given an infinite map, we observe that the set $\mathcal{C}$ of countries is countable, since each country includes at least one point with rational coordinates, and the set of such points is countable. Let $\mathcal{C} = \{C_n : n \in \mathbb{N}\}$.

Take a set $\{p_{n,k} : n \in \mathbb{N}, k = 0, 1, 2, 3\}$ of propositional variables. Now let Σ consist of the following formulae:

(a) for each $n \in \mathbb{N}$, a formula

$$((p_{n,0} \lor p_{n,1} \lor p_{n,2} \lor p_{n,3}) \land ((\neg(p_{n,0} \land p_{n,1})) \land \cdots \land (\neg(p_{n,2} \land p_{n,3})))))$$

asserting that exactly one of $p_{n,0}$, $p_{n,1}$, $p_{n,2}$, $p_{n,3}$ is true;

(b) for each adjacent pair (C_n, C_m) of countries, and each $k \in \{0, 1, 2, 3\}$, the formula

$$(\neg(p_{n,k} \land p_{m,k})).$$

(In (a), we have omitted some brackets for clarity, see Exercise 3.4.)

Now a valuation v satisfying Σ is just a recipe for a legal colouring of the map: assign colour k to country C_n if and only if $v(p_{n,k}) = \mathsf{T}$. The formula (a) guarantees that C_n is given exactly one colour, while (b) shows that neighbouring countries get different colours.

We claim that any finite subset Σ_0 of Σ is satisfiable. For the set of countries C_n for which some $p_{n,k}$ is mentioned in a formula in Σ_0 is finite, and so forms a finite map. By the result of Appel and Haken, it can be coloured. So we can assign values to the relevant $p_{n,k}$ such that $v(\Sigma_0) = \mathsf{T}$. Now assign arbitrary values to the remaining variables.

By the Compactness Theorem, Σ is satisfiable, so the whole map can be coloured. □

The Soundness and Completeness Theorem, and Propositional Compactness, are true more generally:

Theorem 3.8

The conclusions of Theorem 3.4 remain valid for propositional logic in any well-ordered set of propositional variables.

Proof

A formula is a finite string of symbols taken from an alphabet consisting of propositional variables and a finite number of connectives and parentheses. If the set of propositional variables is well-ordered, then the entire alphabet is well-ordered; so the set of all strings is well-ordered, and the subset consisting of formulae is well-ordered (Exercise 2.2).

We follow the same argument as before. The only difficulty occurs in the construction of the maximal consistent set Σ^+. For this, we use transfinite induction. Well-order the set of formulae, as $(\phi_\alpha : \alpha < \beta)$ for some ordinal β. Now the induction proceeds as follows:

- At step zero, take $\Sigma_\alpha = \Sigma$.

- Successor step: if Σ_α has been constructed, set $\Sigma_{s(\alpha)} = \Sigma_\alpha \cup \{\phi_\alpha\}$ or $\Sigma_\alpha \cup \{(\neg\phi_\alpha)\}$, whichever is consistent.

- Limit step: if α is a limit ordinal, let

$$\Sigma_\alpha = \bigcup_{\gamma < \alpha} \Sigma_\gamma.$$

Then take $\Sigma^+ = \Sigma_{s(\beta)}$.

The proof of consistency is as before except at the limit step. But if a contradiction could be derived from Σ_α, where α is a limit ordinal, then only finitely many formulae of this set would be used in the proof, and these formulae would lie in Σ_γ for some $\gamma < \alpha$.   □

The statement that Propositional Compactness holds for *any* infinite set of propositional variables is more problematic; we discuss it briefly in Chapter 6.

3.4 Boolean algebra

The nineteenth century was a time of reductionism and abstraction in mathematics. In geometry, Hilbert revised Euclid's axioms (including the unstated assumptions that Euclid made) and showed that Descartes' description of the Euclidean plane as $\mathbb{R}^2$ is a consequence of these axioms. Group theory, too, moved from descriptive ('a group is a set of transformations with certain closure properties') to axiomatic ('a group is a set with a binary operation satisfying certain axioms') in the hands of Dyck and Cayley.

George Boole's aim was to apply this process and reduce the 'Laws of Thought' to an axiomatic branch of algebra. Indeed, this, or more fully, 'An Investigation of the Laws of Thought, on which are founded the Mathematical Theories of Logic and Probability', was the title of his famous book on the subject. In the introduction to the work, he drew attention to the differences between logic and algebra, and explained that he intended to unify them.

Boole considered those combinations of elementary propositions which we have seen to be described by propositional logic. In order to produce an algebra of propositions, we regard two propositions as the same if they are logically equivalent. Boole based his algebra on the operations $\vee$, $\wedge$ and $\neg$, rather than $\neg$ and $\rightarrow$ as we have done. Perhaps this was because these operations satisfy laws (such as the associative, commutative and distributive laws) which are close to those holding in more familiar algebraic structures such as groups and rings.

We proceed to the definition. A *Boolean algebra* is a set B with two binary operations $\vee$ and $\wedge$, one unary operation $'$, and two distinguished constants 0 and 1, satisfying the following axioms:

- *Associative laws*: $x \vee (y \vee z) = (x \vee y) \vee z$ and $x \wedge (y \wedge z) = (x \wedge y) \wedge z$;

- *Commutative laws*: $x \vee y = y \vee x$ and $x \wedge y = y \wedge x$;

- *Idempotent laws*: $x \vee x = x$ and $x \wedge x = x$;

- *Absorptive laws*: $x \vee (x \wedge y) = x = x \wedge (x \vee y)$;

- *De Morgan's Laws*: $(x \vee y)' = x' \wedge y'$ and $(x \wedge y)' = x' \vee y'$;

- *Identity laws*: $x \vee 0 = x$ and $x \wedge 1 = x$;

- *Complement law*: $x \vee y = 1$ and $x \wedge y = 0$ if and only if $y = x'$.

Clearly not all these laws are independent! From the Identity and Complement Laws, one finds that $0' = 1$ and $1' = 0$. The Complement and Commutative Laws give $(x')' = x$, and then one of De Morgan's laws follows from the other.

Now consider the set of all formulae of propositional logic in a given set of propositional variables. We say that formulae ϕ and ψ are *logically equivalent* if $v(\phi) = v(\psi)$ for all valuations v. This is (as the name suggests) an equivalence relation; let $[\phi]$ denote the equivalence class of the formula ϕ. We define operations on equivalence classes by the rules

$$\begin{aligned} [\phi] \vee [\psi] &= [(\phi \vee \psi)], \\ [\phi] \wedge [\psi] &= [(\phi \wedge \psi)], \end{aligned}$$

$$[\phi]' \;=\; [(\neg\phi)].$$

Further, we let 1 and 0 be the equivalence classes of tautologies and contradictions respectively.

Theorem 3.9

The above operations on the set of equivalence classes of propositional formulae are well-defined; with respect to them this set is a Boolean algebra, denoted by $B(P)$, where P is the set of propositional variables.

The proof involves straightforward verification using truth tables. These algebras represent Boole's 'calculus of propositions'.

Not every Boolean algebra arises in this way, however. Let U be any set, and define operations on $\mathcal{P}\,U$ by the rules

$$
\begin{aligned}
x \vee y &= x \cup y, \\
x \wedge y &= x \cap y, \\
x' &= U \setminus x, \\
1 = U, \qquad & 0 = \varnothing.
\end{aligned}
$$

Again, straightforward verification shows that we have a Boolean algebra. This is Boole's 'algebra of sets'.

The Boolean algebras $B(P)$ (equivalence classes of propositional formulae) can also be represented as subalgebras of those of type $\mathcal{P}\,U$. Consider propositional formulae in a given set P of variables. Let $V(P)$ be the set of valuations. (Any valuation is a function from P to $\{\mathsf{T},\mathsf{F}\}$.) Now, for any formula ϕ, there is an *evaluation function* $e_\phi : V(P) \to \{\mathsf{T},\mathsf{F}\}$, given by

$$e_\phi(v) = v(\phi).$$

By definition, equivalent formulae give the same evaluation function; so equivalence classes of formulae are represented by functions from $V(P)$ to $\{\mathsf{T},\mathsf{F}\}$, or by subsets of $V(P)$ (identifying a function $f : V(P) \to \{\mathsf{T},\mathsf{F}\}$ with the set $\{v \in V(P) : f(v) = \mathsf{T}\}$). Let x_ϕ be the subset corresponding to ϕ. Now

$$
\begin{aligned}
v \in x_{(\phi \vee \psi)} \;&\Leftrightarrow\; v((\phi \vee \psi)) = \mathsf{T} \\
&\Leftrightarrow\; v(\phi) = \mathsf{T} \text{ or } v(\psi) = \mathsf{T} \\
&\Leftrightarrow\; v \in x_\phi \text{ or } v \in x_\psi \\
&\Leftrightarrow\; v \in x_\phi \cup x_\psi,
\end{aligned}
$$

so

$$x_{(\phi \vee \psi)} \;=\; x_\phi \cup x_\psi.$$

Similar calculations show that

$$
\begin{aligned}
x_{(\phi \wedge \psi)} &= x_\phi \cap x_\psi, \\
x_{(\neg \phi)} &= V(P) \setminus x_\phi, \\
x_1 = V(P), \qquad x_0 &= \varnothing.
\end{aligned}
$$

So the map $[\phi] \mapsto x_\phi$ is an isomorphism from the Boolean algebra of equivalence classes of formulae to a subalgebra of the Boolean algebra $\mathcal{P} V(P)$.

This map is not, in general, onto. However, it is onto if P is finite. If $|P| = n$, then $|V(P)| = 2^n$, and so the cardinality of the Boolean algebra is 2^{2^n}. This is readily shown by writing down formulae which are satisfied by any prescribed set of valuations. Let $p_1, \ldots, p_n$ be the propositional variables. Given any valuation v, define a formula τ_v as follows: first, let

$$
q_i = \begin{cases} p_i & \text{if } v(p_i) = \mathsf{T}, \\ (\neg p_i) & \text{if } v(p_i) = \mathsf{F}. \end{cases}
$$

Then set

$$
\tau_v = (q_1 \wedge q_2 \wedge \cdots \wedge q_n).
$$

We see that $v(\tau_v) = \mathsf{T}$ while $v'(\tau_v) = \mathsf{F}$ for any valuation $v' \neq v$. For τ_v can only take the value T if all of its conjuncts do. Now, if x is any non-empty subset of $V(P)$, let

$$
\phi_x = \bigvee_{v \in x} \tau_v
$$

(this notation means the disjunction of all the terms τ_v for $v \in x$). We see that

$$
v(\phi_x) = \mathsf{T} \text{ if and only if } v \in x,
$$

since a valuation satisfies ϕ_x if and only if it satisfies at least one of its disjuncts. So $[\phi_x]$ is mapped to x by our isomorphism. Of course, any contradiction (such as $(p_1 \wedge (\neg p_1)))$ is mapped to the empty set. So the isomorphism is onto the whole of $\mathcal{P} V(P)$.

The formula ϕ_x constructed above to satisfy a given non-empty set of valuations is said to be in *disjunctive normal form*. We summarise our conclusion.

Theorem 3.10

Let P be any set of propositional variables. Then $B(P)$ is isomorphic to a subalgebra of $\mathcal{P} V(P)$. If P is finite, then $B(P) \cong \mathcal{P} V(P)$.

If P is countable, then the set of propositional formulae in the variables P is countable, and so $B(P)$ (the set of equivalence classes of formulae) is countable. Thus it cannot be isomorphic to $\mathcal{P}\, V(P)$, since

$$|\mathcal{P}\, V(P)| > |V(P)| = |\mathcal{P}\, P| > |P|,$$

by Cantor's Theorem.

The axioms for a Boolean algebra, with their commutative, associative, distributive and identity laws, bear some resemblance to those of a ring. The true resemblance is much closer, however. A ring with identity in which every element x satisfies $x^2 = x$ is called a *Boolean ring*, for reasons which the next theorem makes clear.

Theorem 3.11

(a) Let B be a Boolean algebra. Define operations $+$ and $\cdot$ on B by the rules

 - $x + y = (x \vee y) \wedge (x \wedge y)'$,

 - $x \cdot y = x \wedge y$.

 Then $(B, +, \cdot)$ is a Boolean ring, with 0 and 1 as its zero and identity.

(b) Conversely, let R be a Boolean ring. Define new operations $\vee$, $\wedge$, $'$ on R by

 - $x \vee y = x + y + xy$,

 - $x \wedge y = xy$,

 - $x' = 1 + x$,

 and let 0 and 1 be the zero and identity of R. Then R with these operations is a Boolean algebra.

(c) These two constructions are mutually inverse.

Proof

(a) This involves checking that the ring axioms (see Wallace [46], Chapter 3, for example), the identity property of 1, and the fact that $x^2 = x$.

(b) At first the definition of a Boolean ring doesn't seem strong enough. We show how to derive commutativity and the fact that $x + x = 0$ for all x. The rest of the argument is straightforward.

In a Boolean ring, for any x and y, we have

$$\begin{aligned}
x + y &= (x+y)^2 \\
&= x^2 + xy + yx + y^2 \\
&= x + xy + yx + y;
\end{aligned}$$

cancellation gives $xy + yx = 0$, or $yx = -xy$. Specializing to $y = x$, we have $x^2 + x^2 = 0$, or $x + x = 0$, so that $x = -x$. Now this holds for any element x; so

$$yx = -xy = xy,$$

and the multiplication is commutative.

(c) Straightforward.   □

This result fulfils Boole's program, translating propositional logic into ring theory, a very familiar branch of abstract algebra. It is interesting to see how valuations on the Boolean algebra $B(P)$ translate into ring theory. Any valuation v maps the set of formulae to $\{\mathsf{T}, \mathsf{F}\}$; equivalent formulae have the same value, so v induces a map from $B(P)$ to $\{\mathsf{T}, \mathsf{F}\}$. We can identify the elements T and F of the image with the elements 1 and 0 of the Boolean ring $R_2 = \mathbb{Z}/2\mathbb{Z}$ of integers mod 2. This map satisfies $v(1) = 1$, $v(0) = 0$, $v(x+y) = v(x)+v(y)$, and $v(xy) = v(x)v(y)$, so it is a *homomorphism*. Conversely, any homomorphism from $B(P)$ to R_2 arises from a valuation.

EXERCISES

3.1   This exercise gives a decision procedure for Hofstadter's MU-system (a rule for deciding whether or not a formula is a theorem of the system). Prove that a formula is a theorem of the system if and only if it has the properties

• its first letter is M, and all other letters are I or U;

• the number of occurrences of I is not divisible by 3.

3.2   (a) Define a binary propositional connective $\downarrow$ with the following truth table:

ϕ	ψ	$(\phi \downarrow \psi)$
T	T	F
T	F	T
F	T	T
F	F	T

Prove that any propositional formula is logically equivalent to one using only this connective.

(b) Define another binary connective with this property.

(c) Show that, of the sixteen possible binary connectives which could be defined, only two have this property.

3.3   (a) Show that $((\neg\psi) \to (\psi \to \theta))$ is a theorem.

(b) Show that, if both ψ and $(\neg\psi)$ can be deduced from $\Sigma \cup \{(\neg\phi)\}$, then ϕ can be deduced from Σ.

(c) Prove the following theorems:

(i) $(((\neg\phi) \to \psi) \to (((\neg\phi) \to (\neg\psi)) \to \phi))$;

(ii) $((\neg(\neg\phi)) \to \phi)$;

(iii) $((\phi \to \psi) \to ((\neg\psi) \to (\neg\phi)))$

(iv) $((\phi \to \psi) \to (((\neg\phi) \to \psi) \to \psi))$:

(v) $(\psi \to ((\neg\theta) \to (\neg(\psi \to \theta))))$.

3.4   Let $\phi_1, \ldots, \phi_n$ be formulae. Consider the 'pseudo-formula'

$$(\phi_1 \lor \phi_2 \lor \cdots \lor \phi_n).$$

It is possible to insert brackets to make this a well-formed formula in several different ways; for example, if $n = 3$,

$$((\phi_1 \lor \phi_2) \lor \phi_3) \text{ or } (\phi_1 \lor (\phi_2 \lor \phi_3)).$$

Show that, however the brackets are inserted, the truth value of the formula for any valuation is the same. (This justifies our cavalier misuse of brackets in the proof of the Four-Colour Theorem.)

3.5   The following exercise outlines a proof that the Propositional Compactness Theorem, for any set of propositional variables, implies that every set can be totally ordered. Suppose that Propositional Compactness holds in general. Let X be a set, and take a family $\{p_{xy} : x, y \in X, x \neq y\}$ of propositional variables. Now let Σ consist of the following propositional formulae:

- $((p_{xy} \lor p_{yx}) \land (\neg(p_{xy} \land p_{yx})))$, for all distinct $x, y \in X$;

- $((p_{xy} \land p_{yz}) \to p_{xz})$, for all distinct $x, y, z \in X$.

Show that any finite subset of Σ is satisfiable. Show that a valuation v satisfying Σ gives rise to a total ordering of X by the rule that $x < y$ if and only if $v(p_{xy}) = \mathsf{T}$.

3.6   A *graph* consists of a set X of *vertices* with an irreflexive and symmetric relation R of *adjacency* on X. A *colouring* of a graph with a set C of colours is a function f from X to C with the property that, if x and y are adjacent vertices, then $f(x) \neq f(y)$. A *subgraph* of the graph (X, R) is a graph (Y, S), where $Y \subseteq X$ and $S = R \cap Y^2$.

Use the Compactness Theorem to prove that, if (X, R) is a graph whose vertex set X is well-ordered, and if every finite subgraph of the graph (X, R) has a colouring with a given finite set C of colours, then (X, R) has a colouring with this set of colours.

3.7   Prove Theorem 3.11.

3.8   Let $P = \{p_1, \ldots, p_n\}$ be a finite set of propositional variables, $V(P)$ the corresponding set of valuations, so that $|V(P)| = 2^n$. Let $V(P) = \{v_0, \ldots, v_{2^n - 1}\}$.

(a) Consider the set M of formulae which can be built using the connectives $\neg$ and $\leftrightarrow$ only. For any formula $\phi \in M$, let $x_0(\phi)$ be the number of occurrences of $\neg$ in ϕ, and $x_i(\phi)$ be the number of occurrences of p_i in ϕ for $i = 1, \ldots, n$. Prove that a formula $\phi \in M$ is a tautology if and only if $x_i(\phi)$ is even for $i = 0, \ldots, n$. Prove also that two formulae $\phi, \psi \in M$ are logically equivalent if and only if

$$x_i(\phi) \equiv x_i(\psi) \pmod{2}$$

for $i = 0, \ldots, n$.

(b) Prove that, if $\phi \in M$ is not a tautology or a contradiction, then $v(\phi) = \mathsf{T}$ for exactly half of the possible valuations $v \in V(P)$. Deduce that, if $\phi, \psi \in M$ and ψ is not equivalent to ϕ or $(\neg\phi)$, then $v(\phi) = v(\psi)$ for half the possible valuations $v \in V(P)$.

(c) Suppose that a transmitter A wants to send $n + 1$ bits of information $(e_0, \ldots, e_n)$ to a receiver B, over a noisy channel which will introduce some errors in the message. A chooses a formula $\phi \in M$ with $x_i(\phi) \equiv e_i \pmod{2}$, and sends the sequence of 2^n values $v_i(\phi)$ for $v_i \in V(P)$. Show that, provided that fewer than one-quarter of the symbols are transmitted incorrectly, B can recover the information sent by A.

[This error-correction scheme is known as the (first-order) *Reed–Muller code*. Such a code with $n = 5$ was used in one of the early NASA missions to explore the planet Mars.]

3.9   The formal system for propositional logic given in this chapter has infinitely many axioms. Consider the following formal system, which has three axioms and two rules of inference. If ϕ and ψ are formulae and p is a propositional variable, let $\phi[\psi/p]$ denote the formula obtained from ϕ by substituting ψ for every occurrence of p.

The *axioms* are:

(A1) $(p_1 \to (p_2 \to p_1))$

(A2) $((p_1 \to (p_2 \to p_3)) \to ((p_1 \to p_2) \to (p_1 \to p_3)))$

(A3) $(((\neg p_1) \to (\neg p_2)) \to (p_2 \to p_1))$

The *rules of inference* are:

(MP) *Modus Ponens*: From ϕ and $(\phi \to \psi)$, infer ψ.

(S) *Substitution*: From ϕ, infer $\phi[\psi/p]$.

Prove that the Soundness and Completeness Theorem holds for this formal system.

3.10   Prove the Soundness and Completeness Theorem for the following *natural deduction system*: there are no axioms, and three rules of inference:

- *Modus Ponens*: from ϕ and $(\phi \to \psi)$, infer ψ;

- *Contradiction*: from $((\neg \phi) \to \psi)$ and $((\neg \phi) \to (\neg \psi))$, infer ϕ;

- *Deduction Theorem*: if ψ has been inferred from $\Sigma \cup \{\phi\}$, then infer $(\phi \to \psi)$ from Σ.

[How can we prove anything without axioms? For example, we can infer ϕ from $\{\phi\}$, by the definition of a proof from a set of hypotheses; so, using the third rule of inference, we find that $(\phi \to \phi)$ is a theorem.]

3.11   Is there a sound and complete formal system for propositional logic with no rules of inference?

3.12   Logic is often regarded as the foundation on which mathematics is built. Are we justified, then, in using induction on the length of the formula ϕ in the proof of the Deduction Theorem?

4
First-order logic

In this chapter we set up a more complicated formal system, one which is better adapted for the discussion of real mathematics. This is *first-order logic*. It differs from propositional logic in two important ways:

- The basic propositions are no longer indivisible, but make statements about a mathematical structure equipped with relations, functions, and distinguished constants: for example, that two functions, evaluated on specified elements, are equal, or that two elements satisfy a specified relation.

- The language includes quantification, so that we can say that all elements in the structure satisfy a given equation or property, or that some element does.

The development parallels that given for propositional logic in the last chapter: we set up the language, specify its semantics, define a formal system, and prove the Soundness and Completeness Theorem. However, the proof is more complicated, so we defer the discussion of consequences of this theorem to the next chapter.

The term 'first-order logic' is contrasted with various 'higher order logics' (in which we are allowed to quantify over all subsets of the domain, or all relations on the domain, rather than only elements of the domain) and 'infinitary logics' (in which formulae are permitted to involve infinite disjunctions or quantifications).

4.1 Language and syntax

There is not just one 'first-order logic': the system can be adapted to almost any universe of mathematical discourse. For the most part, a system considered in some mathematical subject consists of a set on which various functions and relations are defined, and in which certain particular elements or constants are distinguished. For this purpose, we define an n-ary function (or operation) on a set X to be a function $f : X^n \to X$, and an n-ary relation to be a subset of X^n (which can be thought of as a property which is either true or false for each n-tuple). The terms 1-ary, 2-ary are usually given as *unary* and *binary* respectively. Thus, a binary relation is just a relation on X in the sense defined earlier. A unary relation is simply a subset of X.

For example, a group can be regarded as a set on which we have a binary operation (the group operation or multiplication), a distinguished element (the identity), and a unary operation (inversion, taking each group element to its inverse). Alternatively, we could just take the group operation as basic, and define the identity and inverses in terms of it. A field has two binary operations (addition and multiplication) and two distinguished elements (zero and one); we may also take unary operations for $-x$ and $1/x$. An ordered set, or a graph, has just a single binary relation (for order, or adjacency).

A vector space is a little more complicated. For much of the time we consider vector spaces over a fixed field F. In this case, we can specify the structure by a binary operation (addition) and, for each scalar $c \in F$, a unary operation (multiplication by c). If we want to change the field during the argument, then we must include it in the specification for the vector space. This we do by letting the underlying set consist of both scalars and vectors, and having a unary relation which picks out the scalars. Then we have the appropriate functions to put a field structure on this set, and an abelian group structure on its complement, and a binary function to define scalar multiplication. (As specified, these functions are not everywhere defined; there are various tricks for getting around this which don't concern us now.)

One branch of mathematics which does not easily fit into this framework is topology. A *topological space* consists of a set X and a set $\mathcal{T}$ of subsets of X having certain properties. This cannot be encoded by a fixed number of relations or functions on X. One way round the problem is to work in the more expressive second-order logic which can naturally describe sets of subsets. Another is to express set theory within first-order logic, as we shall do later, in which case the problem disappears! (But this means that topology inherits the perplexities of set theory.)

The language of first-order logic for a particular application will thus contain symbols for the functions, relations and constants appropriate to the application. As usual, these will be nothing but formal symbols; we refer to them as *function symbols, relation symbols* and *constant symbols.* We regard the *arity* of a function or relation symbol as 'built in' to the symbol itself; we say things like 'R is a binary relation symbol'. While these symbols will 'stand for' functions, relations and constants when we define the semantics, it is important not to confuse the function symbol with the function! Some authors distinguish them with different symbols.

The language also contains *variables* for referring to arbitrary elements of the structure. Although formulae will be finite, and so can only contain finitely many variables, we do not want to put any *a priori* upper bound on the number of variables we can use; so we take a countable supply of variables, called $x_0, x_1, x_2, \ldots$. (But if a formula contains only a few variables, we usually call them x, y, z.)

The other ingredients are purely logical symbols: these are of four types:

- *equals sign* $=$.

- *connectives*: We shall permit any of the standard connectives $\neg$ ('not'), $\vee$ ('or'), $\wedge$ ('and'), $\rightarrow$ ('implies'), $\leftrightarrow$ ('if and only if').

- *quantifiers*: There are two of these, $\exists$ ('there exists') and $\forall$ ('for all').

- *punctuation*: left and right parentheses (for making formulae meaningful) and comma (for separating arguments to a function or relation).

These then (the specified function, relation and constant symbols, the standard set of variables, and the equals sign, connectives, quantifiers and punctuation marks) are the symbols of the first-order language.

A *formula* (or *well-formed formula*) will be a string of symbols. However, not all strings will be formulae of the language. We now have to give the specifications for a formula. Remember that we must ensure that it can be tested mechanically whether or not a given string of symbols is a formula or not.

First we define (recursively) a *term* in the language, by the following rules:

- A string consisting just of a variable is a term.

- A string consisting just of a constant symbol is a term.

- If f is an n-ary function symbol and if $t_1, \ldots, t_n$ are terms, then $f(t_1, \ldots, t_n)$ is a term.

As usual, any term must be built by applying these rules. The third rule allows us to build up long terms from shorter ones. For example, if our language contains a binary function symbol f, a unary function symbol g, and constants

c and d, then the expression

$$f(g(c), f(x, f(g(g(f(d, y))), f(c, d)))))$$

is a term.

Next we define an *atomic formula*, by the rules:

- If R is an n-ary relation symbol and if $t_1, \ldots, t_n$ are terms, then $R(t_1, \ldots, t_n)$ is an atomic formula.

- If t_1 and t_2 are terms, then $(t_1 = t_2)$ is an atomic formula.

Finally we define a *formula*, by the rules:

- An atomic formula is a formula.

- If ϕ and ψ are formulae, then each of $(\phi \wedge \psi)$, $(\phi \vee \psi)$, $(\neg\phi)$, $(\phi \to \psi)$ and $(\phi \leftrightarrow \psi)$ is a formula.

- If ϕ is a formula and x a variable, then $(\exists x)\phi$ and $(\forall x)\phi$ are formulae.

Now it is possible to recognize terms and formulae mechanically; a parsing process can be devised, similar to that we saw in propositional logic for recognizing formulae. The parsing can be described by trees, but we have to allow more than two branches from each node; an n-ary function should correspond to a node with n branches in a parsing tree for terms, for example.

We say that ϕ is a *subformula* of ψ if, in the course of the recursive construction of ψ, the formula ϕ appears somewhere. Now the *scope* of a quantifier $\forall$ or $\exists$ in a formula is defined to be the subformula $(\forall x)\phi$ or $(\exists x)\phi$ containing it: this subformula is said to be *quantified over* the variable x. Now an occurrence of a variable x in a formula is said to be *bound* if it is in the scope of a quantifier which is quantified over x. Thus, the x in the $(\forall x)$ or $(\exists x)$ part of the formula, and any occurrence of x in ϕ, is bound. Any occurrence of x which is not bound is said to be *free*. Note that we refer to free and bound *occurrences* of variables, not to free and bound *variables*: the same variable may occur free in one part of a formula and bound in another. We say that a formula is a *sentence* if it has no free occurrences of variables.

For example, suppose that our language contains one binary function symbol μ, one unary function symbol ι, and one constant symbol ϵ. Then the three sequences

$$(\forall x)(\forall y)(\forall z)(\mu(\mu(x, y), z) = \mu(x, \mu(y, z)))$$
$$(\forall x)((\mu(x, \epsilon) = x) \wedge (\mu(\epsilon, x) = x))$$
$$(\forall x)((\mu(x, \iota(x)) = \epsilon) \wedge (\mu(\iota(x), x) = \epsilon))$$

are sentences.

4.2 Semantics

While we were defining the syntax of a first-order language above, we gave informal names to some of the symbols such as 'not', 'if and only if' and 'for all'. These suggest how a formula can be regarded as saying something about a structure. The following definition of the semantics makes this precise.

Let $\mathcal{L}$ be a first-order language. Recall that $\mathcal{L}$ is completely specified by the relation, function and constant symbols it contains: the rest of the symbols are common to all first-order languages. By an $\mathcal{L}$-*structure*, we mean a non-empty set V together with an n-ary relation on V, n-ary function on V, or distinguished element of V, for each n-ary relation symbol, n-ary function symbol, and constant symbol in $\mathcal{L}$, respectively. We denote the actual relations, etc. by the same notation as the relation symbols, etc. of the formal language, but it is important to remember the distinction!

We intend that formulae in the language should say something meaningful about structures over the language. However, the truth of the formula will still depend on the values assigned to the variables. So we say that a *valuation* v of $\mathcal{L}$ consists of an $\mathcal{L}$-structure V together with a sequence $(v_0, v_1, v_2, \ldots)$ of elements of V, these being the values which can be assigned to the variables.

A valuation can be thought of as a mapping of the set of variables into V, where $v(x_i) = v_i$ for all $i \in \mathbb{N}$. We now give rules to extend this to a mapping of the terms of $\mathcal{L}$ into V, and of the formulae of $\mathcal{L}$ into the set $\{T, F\}$ of truth-values. We denote these mappings by the same symbol v. The definitions follow the definitions given above of terms and formulae. Almost all of this involves doing exactly what you would expect!

First, for terms:

- For a variable x_i, $v(x_i) = v_i$ as already explained.

- For a constant symbol c, $v(c)$ is the corresponding element of V.

- If f is an n-ary function symbol and if t_1, ..., t_n are terms, then we let $v(f(t_1, \ldots, t_n))$ be the result of applying the n-ary function on X corresponding to the function symbol f to the arguments $v(t_1), \ldots, v(t_n)$ (which are elements of V already defined).

For atomic formulae:

- If R is an n-ary relation symbol and if t_1, ..., t_n are terms, then we set $v(R(t_1, \ldots, t_n)) = T$ if and only if the relation corresponding to R holds of the n elements $v(t_1), \ldots, v(t_n)$ of V.

- $v((t_1 = t_2)) = T$ if and only if $v(t_1) = v(t_2)$ (as elements of V).

For formulae, we need first another concept. We say that two valuations v and v' are *i-near* if $v(x_j) = v'(x_j)$ for all $j \neq i$: that is, they assign the same values to all the variables except possibly x_i.

- If ϕ is atomic, then $v(\phi)$ is as already defined.

- $v(\phi \vee \psi) = \mathsf{T}$ if and only if $v(\phi) = \mathsf{T}$ or $v(\psi) = \mathsf{T}$.

- $v(\phi \wedge \psi) = \mathsf{T}$ if and only if $v(\phi) = \mathsf{T}$ and $v(\psi) = \mathsf{T}$.

- $v((\neg\phi)) = \mathsf{T}$ if and only if $v(\phi) = \mathsf{F}$.

- $v(\phi \to \psi) = \mathsf{T}$ unless $v(\phi) = \mathsf{T}$ and $v(\psi) = \mathsf{F}$.

- $v(\phi \leftrightarrow \psi) = \mathsf{T}$ if and only if $v(\phi) = v(\psi)$.

- $v((\forall x_i)\phi) = \mathsf{T}$ if and only if, for every valuation v' which is *i-near* to v, we have $v'(\phi) = \mathsf{T}$.

- $v((\exists x_i)\phi) = \mathsf{T}$ if and only if, for some valuation v' which is *i-near* to v, we have $v'(\phi) = \mathsf{T}$.

The second to sixth clauses in this definition simply say that the valuation obeys the familiar truth tables for the five connectives $\vee$, $\wedge$, $\neg$, $\to$ and $\leftrightarrow$, which we saw in the last chapter (Table 3.1). The seventh and eighth clauses, defining the semantics of the quantifiers, are a bit more mysterious. The truth or falsity of a formula depends on the values assigned (by the valuation) to the variables occurring in the formula. Now to assert that $(\forall x)\phi$ is to say, not only that ϕ is true, but that it remains true even if the value assigned to x is arbitrarily changed. And to assert $(\exists x)\phi$ does not require that ϕ is true, only that it can be made true by substituting some suitable value of x for the one given.

We see that $v((\forall x)\phi)$ and $v((\exists x)\phi)$ are independent of the value assigned to x by v. It follows that, for any formula ϕ, the truth-value $v(\phi)$ depends only on the values of the variables which have *free* occurrences in ϕ. In particular, if ϕ is a *sentence* (a formula with no free occurrences of variables), then $v(\phi)$ is entirely independent of the valuation: ϕ is either true or false in the given structure. (Recall that a structure is just a set with appropriate relations, functions and constants, with no values assigned to the variables.)

If the sentence ϕ is true in the structure M, then we write $M \models \phi$, read '*M models ϕ*', or '*M is a model for ϕ*'. The *theory* of a structure M is the set of all sentences which are true in M: we write $\mathrm{Th}(M)$ for this set.

For example, if M is a structure over the first-order language described at the end of the last section, then M is a model for the three sentences given there if and only if M is a group: the functions μ and ι are the group multiplication and inversion, and ϵ is the identity. Note that there are only three sentences given there, translating the associative, identity and inverse laws. Standard

algebra texts (such as Wallace [46], Chapter 3) give a fourth axiom, the closure
law: for all $x, y \in M, \mu(x,y) \in M$. But this is implicit in the semantics of first-
order logic. For μ is interpreted as a binary function on M, that is, a function
from $M \times M$ to M; so the closure law holds.

4.3 Deduction

A formal system for first-order logic will, as usual, consist of three parts: an
alphabet, a set of formulae called *axioms*, and a set of *rules of inference* each
taking a finite set of formulae as input and producing a formula as output.
A *proof* is a finite sequence of formulae, with the property that each formula
in the sequence is either an axiom, or the output of a rule of inference whose
input formulae occur earlier in the sequence. A *theorem* is just the last line of
a proof.

More generally, a *proof of ϕ from a set Σ of formulae* is a sequence of
formulae, each an axiom or a member of Σ or a consequence of earlier formulae
in the sequence using a rule of inference. For example, if Σ is the set of axioms
for a group, we intend that the formulae which can be proved from Σ should
be the theorems of (first-order) group theory.

We will make a small simplification to the language before describing a
formal deduction system for it. The connectives $\neg$ and $\rightarrow$ and the quantifier
$\forall$ are enough to express everything. So we use only these. It is easily checked
from the semantics described in the last section that

- $(\phi \wedge \psi)$ is equivalent to $((\neg\phi) \rightarrow \psi)$;

- $(\phi \vee \psi)$ is equivalent to $(\neg(\phi \rightarrow (\neg\psi)))$;

- $(\phi \leftrightarrow \psi)$ is equivalent to $(\neg((\phi \rightarrow \psi) \rightarrow (\neg(\psi \rightarrow \phi))))$;

- $(\exists x)\phi$ is equivalent to $(\neg(\forall x)(\neg\phi))$.

(We call two formulae 'equivalent' if they have the same truth value in any
valuation in any structure. The first three of these rules are the same as in
Section 3.2.)

We now list the *axioms* of first-order logic. Each statement below is not a
single axiom, but an infinite family, depending on the choice of one or more
formulae, variables and terms. Here α, β, γ denote arbitrary formulae, x a
variable, and t any term. The notation $\phi[t/x]$ means the result of substituting t
for every *free* occurrence of x in ϕ. We adopt the technical convention that such
a substitution is only made if *there is no variable $y \neq x$ such that y occurs in t
and x has a free occurrence in the scope of a quantifier $(\forall y)$ in ϕ.* (If we didn't

require this, then the occurrences of y we introduce by the substitution would unintentionally turn out to be bound.) We separate the axioms into two groups: thoe in the first group are general logical assertions, those in the second refer specifically to the $=$ relation. Each axiom represents a scheme, where ϕ, ψ, θ are formulae, x is a variable, and t, u are terms.

(A1) $(\phi \to (\psi \to \phi))$

(A2) $((\phi \to (\psi \to \theta)) \to ((\phi \to \psi) \to (\phi \to \theta)))$

(A3) $(((\neg\phi) \to (\neg\psi)) \to (\psi \to \phi))$

(A4) $((\forall x)\phi \to \phi[t/x])$

(A5) $((\forall x)(\phi \to \psi) \to (\phi \to (\forall x)\psi))$, if there is no free occurrence of x in ϕ.

(E1) $(t = t)$ for any term t.

(E2) $((t = u) \to (u = t))$ for any terms t, u.

(E3) $((t = u) \to ((u = v) \to (t = v)))$ for any terms t, u, v.

(E4) $((t = u) \to (\phi[t/x, t/y] \to \phi[t/x, u/y]))$.

The first three axioms are identical with the axioms for propositional logic given in the last chapter. Axioms (A4) and (A5) are included to deal with quantification. Axioms (E1)–(E3) will 'mean' that $=$ is interpreted as an equivalence relation. (E4) can be paraphrased, 'If some occurrences of a term are replaced by an equal term in a formula, the resulting formula is logically equivalent to the original'. Curiously, no formula of first-order logic can 'force' the interpretation of $=$ to be actual equality. (Perhaps not so curious, given the puzzlement of philosophers over the question of identity: what rule of logic forces the terms 'Sir Walter Scott' and 'the author of *Waverley*' to be identical?) However, we have already insisted that $=$ is to be interpreted as identity in any first-order structure. This is a non-logical requirement, and some authors do otherwise, by allowing a more general kind of interpretation in which $=$ is not strict equality.

There are just two *rules of inference*:

(R1) (*Modus Ponens*): From ϕ and $\phi \to \psi$, infer ψ.

(R2) (Generalization): From ϕ infer $(\forall x)\phi$, where x is any variable.

In the case of deductions from a set Σ of formulae, there is a restriction which must be observed on the use of (R2): x must not have a free occurrence in any formula in Σ. The intuitive reason for this is clear: ϕ could only be guaranteed to hold for valuations giving x the same value that it takes in the formulae in Σ. In practice, we can usually assume that the set Σ consists of sentences (with no free variables), so that the restriction has no bite. (Typically Σ consists of axioms for some branch of mathematics, and we can suppose that its members are universally quantified over all relevant variables, as we did for the axioms of group theory earlier.)

Note that a proof of ϕ from Σ is exactly the same as a proof of ϕ in the formal system where the formulae of Σ are taken as axioms in addition to the logical axioms (A1)–(A5) and (E1)–(E4), except for one small detail. One of our requirements of a formal system is that there is a mechanical procedure for recognizing the axioms. Not all sets Σ of sentences can be recognized in a mechanical way. This distinction becomes important when we come to consider Gödel's Incompleteness Theorem for arithmetic in the next chapter.

As (A1)–(A3) are the axioms for propositional logic, and *Modus Ponens* is one of the rules of inference, we have 'included' propositional logic within first-order logic. The Propositional Completeness Theorem (Theorem 3.4) has the following consequence:

Theorem 4.1

Let Θ be a tautology of propositional logic in the propositional variables $p_1, \ldots, p_n$, and let $\phi_1, \ldots, \phi_n$ be any first-order formulae. Then the formula Φ obtained from Θ by substituting ϕ_i for p_i (for $i = 1, \ldots, n$) is a theorem of first-order logic.

For the substitution of ϕ_i for p_i throughout the proof of Θ gives a proof of Φ using only (A1)–(A3) and *Modus Ponens*. We call such a formula Φ a *propositional tautology*.

Theorem 4.2 (Deduction Theorem)

Suppose that ϕ can be deduced from a set $\Sigma \cup \{\psi\}$ of formulae. Then $(\psi \to \phi)$ can be deduced from Σ.

Proof

We follow closely the proof of the Deduction Theorem in propositional logic (Theorem 3.3). The cases where ϕ is an axiom, $\phi \in \Sigma$, or $\phi = \psi$, are handled exactly as before. So we may suppose that ϕ follows from earlier steps in the proof using a rule of inference.

The case where ϕ is proved by *Modus Ponens* is handled as before. Finally, suppose that ϕ has the form $(\forall x)\theta$, where x does not occur free in ψ or any formula of Σ, and ϕ has been proved by Generalization from θ. By induction, $(\psi \to \theta)$ can be proved from Σ. Now complete the proof as follows:

$$\Sigma \;\vdash\; (\forall x)(\psi \to \theta)$$
$$\Sigma \;\vdash\; ((\forall x)(\psi \to \theta) \to (\psi \to (\forall x)\theta))$$
$$\Sigma \;\vdash\; (\psi \to (\forall x)\theta)$$

The first formula follows by Generalization from our assumption; the second is (A5), and the third follows by *Modus Ponens*. This completes the proof of the Deduction Theorem.   □

4.4 Soundness and completeness

A formula ϕ in a first-order language $\mathcal{L}$ is said to be *logically valid* if, for any $\mathcal{L}$-structure M and any valuation in M, ϕ receives the value T. If ϕ is a sentence (containing no free variables), then, as we have seen, its truth-value is independent on the valuation, depending only on the structure. So we can say: A sentence ϕ is logically valid if and only if $M \models \phi$ for every $\mathcal{L}$-structure M.

At first, it seems completely out of the question to show that a formula is logically valid: how do we reason about all possible structures? This is exactly what the formal system can do for us. We can show that a formula is not logically valid, by simply exhibiting a structure in which it fails to hold. According to the next result, we can show that a formula is logically valid by simply *proving* it! This is the content of the (first-order) Completeness Theorem. Its converse, the Soundness Theorem, says that any formula which can be proved must be logically valid.

Just as in the propositional case, there is a more general form of the Soundness and Completeness Theorem, asserting that a formula can be deduced from a set Σ of formulae if and only if it is a logical consequence of Σ. Finally, there is a 'consistency' version, asserting that a set of formulae is consistent if and only if it is satisfiable (by some valuation in some $\mathcal{L}$-structure).

We state the theorem for finite or countable languages, as we did for propositional logic. At the end of the chapter, we comment on the general case. As for propositional logic, the reverse implications comprise the Soundness Theorem, the forward implications the Completeness Theorem.

Theorem 4.3 (Soundness and Completeness Theorem)

Let $\mathcal{L}$ be a first-order language with finitely or countably many function, relation and constant symbols.

(a) A formula ϕ is logically valid if and only if it is a theorem.

(b) Let Σ be a set of formulae, ϕ a formula. Then ϕ is a logical consequence of Σ if and only if ϕ can be deduced from Σ.

(c) Let Σ be a set of formulae. Then Σ is consistent if and only if there is an $\mathcal{L}$-structure M and valuation v for which every formula in Σ has the value T.

The proof of the Soundness and Completeness Theorem is fairly long; some details will be merely sketched here.

Proof of Soundness

The reverse implication in (a) is a special case of that in (b) (taking Σ to be empty), so we prove (b). We are given that ϕ is provable from Σ, and have to show that it is true for any valuation in any structure for which the formulae in Σ are true. This is shown by induction on the length of the proof of ϕ from Σ.

If there is a one-line proof, then ϕ is either a member of Σ or an axiom. In the first case, it is true by assumption. Axioms (A1) to (A3) are propositional tautologies. Axioms (A4) and (A5) require reasoning with valuations in structures. Let us consider (A5), for example. Let M be a structure and v a valuation. It suffices, using the truth table for implication, to show that, if $v((\forall x)(\phi \to \psi)) = \mathsf{T}$ and $v(\phi) = \mathsf{T}$, then $v((\forall x)\psi) = \mathsf{T}$. (Recall that x has no free occurrences in ϕ.) So we have to prove that, if v' is any valuation which is i-near to v (where $x = x_i$), then $v'(\psi) = \mathsf{T}$. The assertion $v((\forall x)(\phi \to \psi)) = \mathsf{T}$ shows that $v'(\phi \to \psi) = \mathsf{T}$, so that either $v'(\phi) = \mathsf{F}$ or $v'(\psi) = \mathsf{T}$. But $v'(\phi) = v(\phi) = \mathsf{T}$, since x has no free occurrence in ϕ. So $v'(\psi) = \mathsf{T}$, as required.

The validity of (E1)–(E4) is clear from our interpretation of equality.

The fact that *Modus Ponens* preserves truth follows, as before, from the truth table for implication. Let us consider Generalization. Let $v(\phi) = \mathsf{T}$, and $x = x_i$ a variable not occurring in Σ. Then any valuation v' which is i-near to v agrees with v on the formulae in Σ (since x_i doesn't occur free in Σ), and so also $v'(\phi) = \mathsf{T}$ by the inductive hypothesis. Thus $v((\forall x)\phi) = \mathsf{T}$ by definition.

Now (c) follows from (b): for, if Σ is satisfiable by a valuation v, and Σ is inconsistent, then a contradiction, say ϕ and $(\neg\phi)$, can be deduced from Σ; by (b), $v(\phi) = v((\neg\phi)) = \mathsf{T}$, which is impossible.

Proof of Completeness

We first *claim* that it suffices to prove the forward implication in (c). For suppose that this is true, and let ϕ be a logical consequence of a set Σ. Then the set $\Sigma \cup \{(\neg\phi)\}$ cannot be satisfied (given the value T) by any valuation in any structure. By (c), $\Sigma \cup \{(\neg\phi)\}$ is inconsistent. Using the propositional tautology

$$((\neg p_1) \to (p_1 \to p_2))$$

and *Modus Ponens*, we can deduce anything from it; in particular, we can deduce $(\neg\alpha)$, where α is any instance of an axiom. By the Deduction Theorem, we can deduce $((\neg\phi) \to (\neg\alpha))$ from Σ. Now we have the following deduction from Σ:

$$\Sigma \;\vdash\; ((\neg\phi) \to (\neg\alpha))$$
$$\Sigma \;\vdash\; (((\neg\phi) \to (\neg\alpha)) \to (\alpha \to \phi))$$
$$\Sigma \;\vdash\; (\alpha \to \phi)$$
$$\Sigma \;\vdash\; \alpha$$
$$\Sigma \;\vdash\; \phi$$

So it remains to prove the Claim. This is where the real work lies!

We make a number of reductions and alterations as we go along. Since the language is assumed countable, the set of all formulae is also countable – see Exercise 1.14. We can assume that Σ consists of sentences, by appropriate universal quantification.

Step 1

We first enlarge $\mathcal{L}$ by adjoining an infinite sequence $(c_0, c_1, c_2, \ldots)$ of constant symbols which don't already appear. The language is still countable. The role of these mysterious new constants is, roughly speaking, to act as 'witnesses' to the truth of existential sentences.

We make a list of all the formulae of $\mathcal{L}$ which involve just one *free variable* (variable having a free occurrence), say $\alpha_0, \alpha_1, \ldots$, where x_{i_m} is the variable occurring free in α_m. We choose a subsequence $(d_0, d_1, \ldots)$ of the sequence $(c_0, c_1, \ldots)$ of new constant symbols in such a way that the terms of the sequence are all distinct and d_n does not occur in α_m for any $m \leq n$. We now let θ_n be the sentence

$$((\neg(\forall x_{i_n})\alpha_n) \to (\neg\alpha_n[d_n/x_{i_n}])).$$

Informally, if $(\exists x)\phi$ is consistent, with $\phi = (\neg\alpha_n)$ and $x = x_{i_n}$, then d_n witnesses the fact, in that $\phi[d_n/x]$ will end up true in our structure.

We adjoin the sentences θ_n to Σ, and show that the result T is still consistent. It is enough to prove this when one of these sentences is adjoined. Now suppose that a contradiction could be derived after adding the sentence

$$((\neg(\forall x_{i_n})\alpha_n) \to (\neg\alpha_n[d_n/x_{i_n}]))$$

to Σ. Let x be a new variable not occurring anywhere in the proof of the contradiction. Using Generalization and the Deduction Theorem, we can prove

$$(\alpha_n[x/x_{i_n}] \to (\forall x_{i_n})\alpha_n),$$

and hence (using a propositional tautology),

$$((\neg(\forall x_{i_n})\alpha_n) \to (\neg\alpha_n[x/x_{i_n}])).$$

Now take the rest of the proof of the contradiction, and replace d_n by x throughout. We obtain a proof of a contradiction from Σ, contrary to assumption.

The point here is quite subtle. Raymond Smullyan [43] introduces it by telling the story of a man who walks into a bar and tells the barman to set up drinks for everyone, 'Because, when I drink, everybody drinks.' After a few rounds in this vein, the other customers are horrified when the man says, 'And when I pay, everybody pays.' Smullyan remarks that there is a man with the property that, if he drinks, then everybody drinks. For either it is true that everybody drinks, in which case anyone could be used as the witness; or it is not true, in which case there is a man who doesn't drink, and he can be taken to be the witness.

Step 2

Now we use the following result, which is analogous to the key step in the proof of the Propositional Compactness Theorem (and is proved in exactly the same way). A set Σ is *complete* if, for any formula ϕ, either ϕ or $(\neg\phi)$ is in Σ.

Lemma 4.1

(a) If Σ is a consistent set of formulae and ϕ a formula, then either $\Sigma \cup \{\phi\}$ or $\Sigma \cup \{(\neg\phi)\}$ is consistent.

(b) Any consistent set Σ is contained in a complete consistent set.

Part (b) is proved by listing all formulae in a sequence $(\phi_0, \phi_1, \ldots)$, and at the nth stage, adjoining either ϕ_n or $(\neg\phi_n)$ to the set, whichever is consistent.

We use this result to enlarge the consistent set T to a complete consistent set T^+.

Step 3

We use T^+ to construct a structure satisfying the original set Σ. This must be constructed out of the formal system with our bare hands!

The underlying set V of the structure M is taken to be the set of all *closed terms* of the language, terms containing no variables at all. Each constant symbol is a closed term; we interpret it as itself (regarded as an element of the model). In particular, all the witnesses belong to V.

Given an n-ary function symbol f and n closed terms $t_1, \ldots, t_n$, we say that the result of applying the corresponding function to $t_1, \ldots, t_n$ (thought of as elements of V) is $f(t_1, \ldots, t_n)$ (this is a closed term, hence is in V).

Now let R be an n-ary relation symbol and let $t_1, \ldots, t_n$ be closed terms. We declare that $R(t_1, \ldots, t_t)$ is true if and only if this sentence is deducible from T.

We have now defined a structure M.

Step 4

The structure M defined in this way is not quite the one we want, however. It may be provable from T that two closed terms are necessarily equal. (For example, there might be a sentence $(c = c')$ in Σ equating two constant symbols.) So we have to shrink M a bit. Define a relation $\sim$ on V by the rule that $t_1 \sim t_2$ if and only if the sentence $(t_1 = t_2)$ is provable from T. It follows easily from axioms (E1)–(E3) that $\sim$ is an equivalence relation on V. The actual structure we require is the set of equivalence classes of this relation; it is easy to see that the relations and functions on V induce relations and functions on this 'quotient' $\overline{M} = M/\sim$.

It remains to show that $\overline{M}$ is a model of Σ. This requires detailed checking and is not given here.

This step is related to the fact that we cannot specify by any set of first-order sentences that the interpretation of a binary relation is equality: the best we can do is to force it to be an equivalence relation satisfying Leibniz' principle (E4). $\square$

The case we have considered, where the language is countable, is sufficient for many mathematical purposes (such as group theory and the theory of the natural numbers). There are some situations for which it will not do. For example, given a field F, the first-order theory of vector spaces over F requires a unary function symbol for each $c \in F$ (to represent scalar multiplication by c), and so the language is uncountable if the field is: this includes the case of the field $\mathbb{R}$, as we have seen.

The only places in the proof where we used the countability assumption were in Steps 1 and 2, where we arranged formulae of some kind into an infinite sequence and applied an inductive construction to them. Everything will work in the same way if we arrange the formulae in a well-ordered sequence and use transfinite induction, as we did in the corresponding proof for propositional logic (Theorem 3.8).

So it will be important for such applications to know whether a given set can be well-ordered. We will look at this in our discussion of the upward Löwenheim–Skolem Theorem in Chapter 5, and the Axiom of Choice in Chapter 6. For now, note that the Soundness and Completeness Theorem remains valid if we replace the assumption '$\mathcal{L}$ is countable' with the assumption '$\mathcal{L}$ is well-ordered'. (Exercise 2.2 shows that, if the set of symbols in the language is well-ordered, then the set of finite strings of symbols is also well-ordered, and so the set of formulae is well-ordered.) We state this result explicitly for reference.

Theorem 4.4

The conclusions of Theorem 4.3 remain valid over a language $\mathcal{L}$ in which the sets of relation symbols, function symbols and constant symbols can be well-ordered.

EXERCISES

4.1 Let G denote the set of group axioms given at the end of Section 4.1. Show that $G \vdash \sigma$, where σ is the sentence

$$((\forall x)(\mu(x,x) = \epsilon) \to (\forall x)(\forall y)(\mu(x,y) = \mu(y,x)).$$

Hint: It is probably easier to prove this in the 'language of mathematics' first and translate the proof into the first-order language. The point is that mathematical proofs can be written in this language, even though they are somewhat clumsy; and, in this form, checking their correctness is a purely mechanical procedure.

4.2 For each of the cases (a) fields, (b) totally ordered sets, (c) graphs, give a first-order language and a set of sentences which axiomatizes the relevant class of structures.

Can this be done for (d) topological spaces, (e) well-ordered sets?

4.3 Can you suggest any reasons why algebra textbooks insist on the closure law as one of the axioms for a group, instead of allowing it to be implicit in the statement that the group operation is a binary function?

5
Model theory

Model theory is the study of the relationship between formal systems and the mathematical structures they are designed to describe.

Not every formal system is designed to describe mathematical structures: Hofstadter's MU-system is purely didactic, and propositional (Boolean) logic is designed just to describe the truth or logical validity of compound propositions. But first-order logic provides a description of most of mathematics.

More specifically then, first-order model theory is the study of the relationship between sets of sentences in a first-order language and their models. This just moves to a higher level of abstraction the ordinary processes of mathematics, which study the models of a *particular* set of sentences (for example, the axioms of group theory).

There are two broadly different situations. One is exemplified by the group axioms, whose richness and power derives from the wide variety of models they have: a general theorem of group theory applies to the whole diverse collection of groups. At the other extreme, we have Euclid's axioms for geometry, which were at least intended to describe the one true geometry.

5.1 Compactness and Löwenheim–Skolem

The Compactness Theorem and the Löwenheim–Skolem Theorem are two of the most fundamental facts in first-order model theory, on which much else depends. They both follow from Gödel's Completeness Theorem, outlined in the last chapter, and its proof. The Compactness Theorem resembles the propositional version mentioned in the Appendix to the last chapter. Throughout this chapter, we work over a *countable* first-order language.

Theorem 5.1 (Compactness Theorem)

If a set Σ of sentences in a countable first-order language has the property that every finite subset of Σ has a model, then Σ has a model.

Theorem 5.2 (Löwenheim–Skolem Theorem)

If a set Σ of sentences in a countable first-order language has a model, then it has a model which is at most countable (that is, finite or countable).

Proof of the Compactness Theorem

According to the Completeness Theorem, a set Σ of sentences has a model if and only if it is *consistent* (that is, no contradiction can be derived from it). Now suppose that Σ is a set of sentences which has no model. Then there is a proof of a contradiction from the set Σ of hypotheses. Each formula in the proof is an axiom, a member of Σ, or a consequence of earlier formulae in the list using a rule of inference. Since a proof is a finite sequence of formulae, only finitely many members of Σ can occur in it; that is, we have a proof of a contradiction from a finite subset of Σ. Necessarily this finite subset has no model. □

Proof of the Löwenheim–Skolem Theorem

This follows, not from the Completeness Theorem, but from its method of proof. Suppose that Σ is a set of sentences which has a model. Then Σ is consistent. In the proof of the Completeness Theorem, there are three steps relevant here:

- We add a countable set of new constant symbols to the language, and enlarge Σ to a complete set T. Note that the new language and T are still countable.

- We construct a structure N whose elements are the closed terms of the language. Clearly N is countable.

- We define an equivalence relation $\sim$ on N by setting $t_1 \sim t_2$ if the sentence $(t_1 = t_2)$ is provable from T. Now we build a structure N on the set of equivalence classes of this relation. Then N is the required model of T (and so of Σ); and N is at most countable. □

In fact, a more general version of both theorems holds. We can weaken the requirement of a countable language and ask only that the symbols in the language can be well-ordered. (It is enough to require that the sets of relation, function and constant symbols can be well-ordered.) We use the Completeness Theorem in the same way, with transfinite induction replacing ordinary induction in the proof.

Both theorems have numerous applications. The Compactness Theorem implies the following result, the Upward Löwenheim–Skolem Theorem, which asserts that if a set of sentences has an infinite model, then it has arbitrarily large models.

Theorem 5.3 (Upward Löwenheim–Skolem Theorem)

Let Σ be a set of sentences and X any infinite well-ordered set. Suppose that Σ has an infinite model. Then Σ has a model M such that there is an injective map from X to M.

After our discussion of cardinality in the next chapter, we will be able to express this more simply: if Σ has an infinite model, then it has a model of arbitrarily large cardinality. In fact, since the ordinals 'continue for ever', this shows that there are infinitely many different models of Σ (if there is at least one infinite model).

Proof

We extend the language with a set $\{c(x) : x \in X\}$ of constant symbols in one-to-one correspondence with X, and add to Σ the set T consisting of all the sentences $(c(x) \neq c(y))$ for all distinct $x, y \in X$. Now let N be the given model of Σ. We can interpret the new constant symbols in N in such a way that any given finite set of the new sentences are all true. (Since N is infinite, it is large enough that we can assign distinct elements of it to all the constant symbols $c(x)$ mentioned in the sentences.) So every finite subset of $\Sigma \cup T$ has a model. By the Compactness Theorem, $\Sigma \cup T$ has a model, say M. Then clearly M is a model of Σ, and the map $x \mapsto c(x)$ is an injection from X to M, as required. $\square$

As another application, here is a 'first-order' proof that the infinite Four-Colour Theorem follows from the finite. Definitions are given in Chapter 3.

Theorem 5.4

Any plane map can be coloured with four colours.

Proof

Let a map M be given. We note first that the set of countries is at most countable, since each country contains a point of the plane with rational coordinates. We consider the following first-order language L: there is a constant symbol c_i for each country C_i; and there are four unary relations K_1, K_2, K_3, K_4. Let Σ consist of the following sentences:

- For each country C_i, a sentence asserting that exactly one of $K_1(c_i)$, $K_2(c_i)$, $K_3(c_i)$, $K_4(c_i)$ holds;

- For each pair C_i and C_j of countries sharing a non-trivial border, and each $k \in \{1, 2, 3, 4\}$, the sentence $\neg(K_k(c_i) \wedge K_k(c_j))$.

Now a model for Σ corresponds to a colouring of the map, by the rule that C_i is given the kth colour if and only if $K_k(c_i)$ holds in the model. The sentences ensure that each country has a colour, and neighbouring countries have different colours.

Consider any finite subset Σ_0 of Σ. The sentences in Σ_0 only involve a finite set of the constant symbols c_i. Let M_0 be the map consisting of the countries C_i for which c_i occurs in Σ_0. Then M_0 can be coloured with four colours, by the Four-Colour Theorem; so Σ_0 has a model. By the Compactness Theorem, Σ has a model; so M can be coloured with four colours.   □

The Löwenheim–Skolem Theorem has implications of a different sort. It shows, for example, that there is no set of first-order axioms whose only model is the field $\mathbb{R}$ of real numbers, or Euclidean plane geometry: for these axioms will necessarily have a countable model. We will see a more dramatic instance of this (the *Skolem paradox*) in the next chapter.

5.2 Categoricity

We now consider the question: What sort of structures can be described completely by first-order sentences?

First, we have to say what we mean by this. An *isomorphism* between two L-structures M and N is a bijection ϕ from the underlying set of M to that of N which 'preserves the structure', in the following sense:

- For any n-ary relation symbol R and any $a_1, \ldots, a_n \in M$, if $R(a_1, \ldots, a_n)$ holds in M, then $R(\phi(a_1), \ldots, \phi(a_n))$ holds in N.

- For any n-ary function symbol f and any $a_1, \ldots, a_n \in M$,

$$f(\phi(a_1), \ldots, \phi(a_n)) = \phi(f(a_1, \ldots, a_n)).$$

- For any constant symbol c, $\phi(c) = c$.

The last confusing condition means that ϕ maps the element of M interpreting c to the element of N interpreting c. This is common mathematical practice, as when we say that a ring homomorphism maps 0 to 0 (but the zeros lie in different rings). A similar interpretation has to be made in the other cases.

As always in mathematics, we regard isomorphic structures as being 'essentially the same'. If any two structures having property P are isomorphic, we say that property P characterizes the structure in question.

If M is finite, we can say everything about it: we can name its elements, and list all instances of relations they satisfy and the values of all functions on all choices of arguments. If the language contains only finitely many non-logical symbols, then it is characterized by a finite set of sentences (and in fact, by taking the conjunction of all these sentences, we can do it with just one sentence).

However, if M is infinite, we cannot describe it completely by any set of first-order sentences, even by an infinite set. (This follows from the Upward Löwenheim–Skolem Theorem: we saw that the theory of M has infinitely many, arbitrarily large, models.) The best we can hope for is the following.

We say that the structure M, over a first-order language L, is *countably categorical* or *ω-categorical* if any countable structure over L which satisfies the same first-order sentences as M is isomorphic to M.

Informally, M is countably categorical if there is a set of first-order 'axioms' such that any countable model of these axioms is isomorphic to M.

Cantor's Theorem on $\mathbb{Q}$ is a well-known example:

Theorem 5.5

The ordered set $\mathbb{Q}$ is countably categorical. More precisely, if a countable totally ordered set $(X, <)$ is *dense* (that is, between any two distinct elements of X we can find another) and has no greatest or least element, then it is isomorphic to $\mathbb{Q}$.

We outlined a proof of this theorem in Exercise 1.16.

Rather unexpectedly, it turns out that countable categoricity is a symmetry condition. Let M be an L-structure on the set X. An *automorphism* of M is an isomorphism from M to itself: that is, a permutation of X which is an isomorphism from M to M. The set of automorphisms of M forms a group of permutations of X, the *automorphism group* of M (written Aut(M)).

Any permutation group G on X divides the set X into *orbits*: two points are in the same orbit if some permutation in G carries one to the other. (This is an equivalence relation, whose equivalence classes are the orbits: see Exercise 1.5.) Also, any permutation group on X has an induced action on X^n for each positive integer n: it acts coordinatewise on the n-tuples. We say that the permutation group G on X is *oligomorphic* if, for each n, the number of orbits on X^n is finite. Now we have the following:

Theorem 5.6 (Theorem of Engeler, Ryll-Nardzewski and Svenonius)

The countable structure M is countably categorical if and only if its automorphism group is oligomorphic.

We will not attempt to describe the proof of this theorem. The most difficult part resembles the proof of the Completeness Theorem, with an extra difficulty: if Aut(M) is not oligomorphic, we have to construct two non-isomorphic countable models of the theory of M.

We illustrate the theorem by reconsidering $\mathbb{Q}$ (as ordered set). Any two intervals in $\mathbb{Q}$ are isomorphic: we can map (a, b) to (c, d) by the linear function $x \mapsto c + (d - c)(x - a)/(b - a)$. Similarly, any two 'semi-infinite' intervals $(-\infty, a)$ and $(-\infty, b)$ are isomorphic, by means of a translation, and similarly for intervals (a, ∞).

Given two n-tuples $(a_1, \ldots, a_n)$ and $(b_1, \ldots, b_n)$ of distinct rationals, we can re-order them so that $a_1 < \ldots < a_n$ and $b_1 < \ldots < b_n$. Then we can map each interval (a_i, a_{i+1}) to (b_i, b_{i+1}), the end $(-\infty, a_1)$ to $(-\infty, b_1)$, and (a_n, ∞) to (b_n, ∞). The compound map defined in this way (a piecewise-linear function) is an automorphism of $\mathbb{Q}$ which maps the first n-tuple to the second. Similarly, each re-ordering of an n-tuple determines a unique orbit. It follows that Aut($\mathbb{Q}$) has exactly $n!$ orbits on ordered n-tuples of distinct elements, and hence only a finite number on arbitrary n-tuples. So Aut($\mathbb{Q}$) is oligomorphic. This is in agreement with the theorem, since we already know that $\mathbb{Q}$ is countably categorical (from Cantor's characterization).

5.3 Peano arithmetic

A set Σ of first-order sentences is *complete* if, for every sentence α, either α or
$(\neg\alpha)$ is a logical consequence of Σ. This is a usage of the term which is different
from that in the Completeness Theorem (where it is the formal system which
is said to be complete); but it is very closely related, since if Σ is complete
then, for any α, either α or $(\neg\alpha)$ can be proved from Σ.

The *theory* of a structure M is the set of all first-order sentences σ which
hold in M (that is, for which $M \models \sigma$). As we have observed, the theory of M
is necessarily complete.

A given set Σ of sentences which has an infinite model necessarily has more
than one model. But, if Σ is complete, then all its models have the same first-
order theory. This is a relation weaker than isomorphism, which is known by
the term *equivalence*.

Suppose, then, that we set out to axiomatize some area of mathematics. If
the area is like group theory, we do not want the axioms to form a complete
set, since we must allow many different models. On the other hand, if we are
considering an object of particular importance, it would be natural to take a
complete set of sentences, in order to pin it down as far as we can within first-
order logic. (If, for example, the structure is countably categorical, we could
take a complete set of first-order axioms and add the non-logical requirement
of countability.)

A particularly interesting case study is provided by the natural numbers.
These are the most basic objects in mathematics, and the first that we meet,
when as children we learn to count.

One way to proceed is as follows. We have seen in Chapter 2 that the
natural numbers can be constructed, and the addition, multiplication, and order
relation defined, as particular sets. (The natural numbers are the finite ordinals,
that is, all those which can be obtained from zero by the successor operation
only, without taking limits.) In the next chapter, we will give axioms for set
theory, and thereby implicitly axiomatize the natural numbers (within a more
powerful theory).

Could we axiomatize $\mathbb{N}$ directly, without getting it as a by-product of set
theory? A system of axioms should reflect the way in which the natural numbers appear: starting from zero, any number can be obtained by finitely many
applications of the successor function which takes us from each number to the
next. (That is, in principle we can count up to any number.) But we cannot
talk about 'an arbitrary finite number of steps': this would require a formula
along the lines

$$(\forall x)(\exists x_1)\ldots(\exists x_{n-1})(x_1 = s(0) \wedge \ldots \wedge x = s(x_{n-1})),$$

and such formulae are not permitted in first-order logic.

We note in passing that $\mathbb{N}$ has no non-trivial automorphisms at all. For any automorphism fixes 0, and hence fixes 1, and hence ... (The dots, as usual, conceal a proof by induction. In Chapter 2 we showed the more general result that any ordinal has no non-trivial automorphisms.) Hence $\mathbb{N}$ is not oligomorphic, so we know in advance that its theory will have countable models which are not isomorphic to $\mathbb{N}$. The best we can hope for is to give a simple complete set of axioms, from which all properties of $\mathbb{N}$ can be deduced.

The succession principle is the basis of induction, and an alternative way to proceed is to require that the Principle of Induction (in its usual form) is valid. Now the problem is this. We want to say that if a property P holds for 0, and if its truth for n implies its truth for $n + 1$, then it holds for all n. But first-order logic only allows us to talk about properties expressible in first-order logic. (But for this limitation there would be no problem: we could let P be the property 'n is a natural number'.)

We need to begin with a couple of axioms for the successor function. Our language contains a constant symbol 0 and a unary function s (the *successor function*).

(P1) Every element x except zero is the successor of a unique y.

(P2) 0 is not the successor of anything.

These are easily written as first-order sentences. For example, (P2) takes the form $(\forall x)(s(x) \neq 0)$ or, more precisely, $(\forall x)(\neg(s(x) = 0))$.

Note that already we see that (P1) and (P2) have infinitely many non-isomorphic countable models (see Exercise 5.2).

We want to be able to handle formulae referring to addition and multiplication, such as *Goldbach's Conjecture*, the assertion (yet unproved) that every even number greater than 2 is the sum of two primes:

$$(\forall x)(((x \neq 0) \wedge (x \neq 1)) \rightarrow (\exists y)(\exists z)(P(y) \wedge P(z) \wedge (2 \cdot x = y + z))),$$

where 1 and 2 stand for $s(0)$ and $s(s(0))$, and $P(y)$ ('y is prime') is an abbreviation for

$$((y \neq 0) \wedge (y \neq 1) \wedge (\forall u)(\forall v)((y = u \cdot v) \rightarrow ((u = 1) \vee (v = 1)))).$$

We emphasise that, although addition and multiplication can be defined in terms of the successor function, we need names for them in the language in order to write such formulae.

So we define addition and multiplication by induction:

(P3) $(\forall x)(x + 0 = x)$

(P4) $(\forall x)(\forall y)(x + s(y) = s(x + y))$.

(P5) $(\forall x)(x \cdot 0 = 0)$

(P6) $(\forall x)(\forall y)(x \cdot s(y) = x \cdot y + x)$.

Finally, the Principle of Induction, which is not a single axiom, but an *axiom schema*, consisting of one axiom for each formula ϕ in our language with just a single free variable x:

(P7) $((\phi[0/x] \wedge (\forall x)(\phi \to \phi[s(x)/x])) \to (\forall x)\phi)$.

We call this system *Peano arithmetic*. It is not countably categorical: it has models other than $\mathbb{N}$ (such models are called 'non-standard models'). Here is a direct proof that non-standard models of Peano's axioms exist, which suggests what they look like. We adjoin to the language a new constant symbol c. Let σ_n be the sentence $(c > n)$ in the new language, where, as usual, n is short for $s(s(\ldots s(0)\ldots))$, with n occurrences of s. Let

$$\Sigma = \{\sigma_n : n \in \mathbb{N}\}.$$

Then any finite subset Σ_0 of Σ is consistent with Peano's axioms, since we may interpret c as a natural number greater than the largest index of any formula σ_n in Σ_0. By the Compactness Theorem, there is a model of Peano's axioms in which Σ is satisfied. In this model, c is interpreted as an 'infinite number'.

Is Peano arithmetic complete? The remarkable fact that it is not (and so there is no hope of proving all true statements about $\mathbb{N}$ by formal methods, destroying Hilbert's program) was proved by Kurt Gödel in 1930, when he showed that this axiom system is not complete. In other words, there is a sentence σ which is true in $\mathbb{N}$ but not provable from Peano's axioms.

Gödel's Theorem is one of the great intellectual achievements of the twentieth century. We outline the proof, omitting some detailed verifications along the way.

Gödel's big idea was to encode formulae by natural numbers. This procedure is known as *Gödel numbering*. A natural number, in our usual base 10 representation, is just a string of symbols from the alphabet $\{0, 1, 2, 3, 4, 5, 6, 7, 8, 9\}$; a formula is a string of symbols from a different alphabet, so it is not too surprising that we can build a function from the set of formulae to the set of integers. It is not quite straightforward, because the alphabet for first-order arithmetic is infinite: it has infinitely many variables. There are various tricks for getting around this. Usually, we use the Fundamental Theorem of Arithmetic: a natural number can be written uniquely as a product of primes (up to order), and there are infinitely many primes. The version here is simpler, and is based on Hofstadter's 'Typographic Number Theory' [22].

First we encode the symbols of the language (other than the variables) as digits, as shown in Table 5.1.

¬	→	∀	(	)	=	s	+	·	0
1	2	3	4	5	6	7	8	9	0

Table 5.1. Encoding symbols

We will need two further symbols, a 'variable marker' and a 'formula marker', which we will denote by A and B. We are thus required to do our arithmetic to the base 12 instead of 10. However, this is only a minor annoyance, and does not really affect the argument.

The job of the 'variable marker' is something like that of <VAR> and </VAR> in the computer language HTML, with one difference. We will interpret a string of digits between two As as the index of a variable, not the actual variable name. Thus, $A2A$ will refer to the variable x_2. Now to avoid parsing problems, we will take the digit string as the base 10 representation of the variable index, so that A is not used: thus $A35A$ refers to x_{35}, not x_{41}.

The 'formula marker' is not needed just yet. Now any formula can be translated into a string of digits; this string is the base 12 representation of a natural number, which we take to be the *Gödel number* of the formula.

For example, the sentence

$$(\forall x_0)(\neg(s(x_0) = 0))$$

(one of the Peano axioms) is represented by the integer

$$43A0A541474A0A56055$$

in base 12 (which happens to be 115011419876663965121 in base 10). Even quite simple formulae translate into very large numbers.

Now we define the *Gödel number* of a sequence of formulae by terminating each formula with the 'formula marker' B (rather like the semicolon in the programming language C), and just concatenating the results and reading the string as a number to base 12. In particular, any proof has a Gödel number.

We denote by $G(\phi)$ the Gödel number of the formula ϕ.

Recall that a fixed natural number n can be represented in the first-order language by the term $s(s(\ldots s(0) \ldots))$. So, for example, 2 is $s(s(0))$. If we write a natural number in a formula, it is assumed that this representation is meant.

The key to Gödel's proof is the following result.

Theorem 5.7

(a) There is a formula π of first-order arithmetic, with two free variables x_0 and x_1, such that

- $\mathbb{N} \models \pi[n/x_0, m/x_1]$ if $m = G(\phi)$ and n is the Gödel number of a proof of ϕ in PA,

- $\mathbb{N} \models (\neg\pi[n/x_0, m/x_1])$ otherwise.

(b) There is a formula ω of first-order arithmetic, with two free variables x_0 and x_1, such that

- $\mathbb{N} \models \omega[m/x_0, n/x_1]$ if $m = G(\phi)$ for some formula ϕ in which the variable x_0 has a free occurrence, and n is the Gödel number of a proof of $\phi[m/x_0]$ from PA,

- $\mathbb{N} \models (\neg\omega[m/x_0, n/x_1])$ otherwise.

In a sense, the devil is in the details. This theorem is at the end of a sequence of similar results showing that various properties of the Gödel numbering are expressible in a similar way by formulae: for example, that $n = G(\phi)$ for some formula ϕ, that $n = G(\phi)$ for some axiom ϕ, and so on. More precisely, any property of a string of symbols which can be verified mechanically is represented by a formula, the formula being true when evaluated at n if and only if n is the Gödel number of the string. Moreover, the operation of substitution of a formula for a variable is also mechanical, and can be described by a formula in the same way. Part (b) then follows from (a). We do not give the details here.

Gödel's Theorem states that, if Peano arithmetic is consistent, then there are statements which are neither provable nor disprovable. For simplicity, we prove something a bit weaker. We say that Peano arithmetic is *ω-consistent* if, whenever ϕ is a formula with free variable x such that $\phi[n/x]$ is provable for each natural number n, the sentence $(\neg(\forall x)\phi)$ is not provable. This condition is stronger than consistency: for there certainly exist formulae ϕ for which $\phi[n/x]$ is provable for all n (an example is $(x = x)$); then ω-consistency implies that at least some formula (namely $(\neg(\forall x)\phi)$) is not provable. But, if Peano arithmetic were inconsistent, everything would be provable!

If it were possible to give first-order sentences having $\mathbb{N}$ as the only model, then of course ω-consistency would be the same as consistency. But, as we have seen, this hope is unrealizable.

We shall show that, if Peano arithmetic is ω-consistent, then there is a formula which is neither provable nor disprovable. In fact, consistency is enough, but the argument is a little more elaborate.

Now we reach the heart of the mystery, and find self-reference. Let p be the Gödel number of the formula

$$(\forall x_1)(\neg\omega).$$

(Remember that ω has free variables x_0 and x_1.) Let ζ be obtained by substituting p for x_0 in this formula. Thus, the interpretation of ζ is that, for every

n, it is not the case that n is the Gödel number of a proof of the formula $\psi[p/x_0]$, where ψ has Gödel number p. But we know that the formula ψ with this property is $(\forall x_1)(\neg\omega)$, and the result of substituting p for x_0 is precisely ζ. Thus, the interpretation of ζ is the unprovability of ζ.

Now we show that, if Peano arithmetic is ω-consistent, then ζ can be neither proved nor disproved.

Assume that Peano arithmetic is consistent. Then ζ is unprovable. For let it be provable, and let q be the Gödel number of a proof. Setting ϕ to be $(\forall x_1)(\neg\omega)$, we see that

$$\text{PA} \vdash \omega[p/x_0, q/x_1].$$

But also, by assumption, $(\forall x_1)(\neg\omega[p/x_0])$ is provable (this is just ζ); by (A5) and *Modus Ponens*, we get

$$\text{PA} \vdash (\neg\omega[p/x_0, q/x_1]).$$

So Peano arithmetic is inconsistent, contrary to assumption.

Assume now that Peano arithmetic is ω-consistent. Then $(\neg\zeta)$ is not provable. For suppose that it is. Since $(\neg\zeta)$ is the formula

$$(\neg(\forall x_1)(\neg\omega[p/x_0])),$$

it follows that $(\neg\omega[p/x_0, q/x_1])$ is not provable, for some q. By the definition of ω, this means that q is the Gödel number of a proof of $\phi[p/x_0]$, where p is the Gödel number of ϕ. By definition of p, we see that ϕ is $(\forall x_1)(\neg\omega)$, so that $\phi[p/x_0]$ is ζ: but we know that there is no proof of ζ.

As noted, it is possible to strengthen this conclusion.

Theorem 5.8 (Gödel's Incompleteness Theorem)

If Peano arithmetic is consistent, then there is a sentence which is neither provable nor disprovable in Peano arithmetic.

The sentence in question, like ζ in the above proof, asserts its own unprovability (more precisely, that there is no natural number which is the Gödel number of a proof of it). So, in $\mathbb{N}$ it is true. This shows Tarski's variant of Gödel's Theorem:

Theorem 5.9 (Tarski's Theorem)

If Peano arithmetic is consistent, then there is a sentence which is true in $\mathbb{N}$ but not provable in Peano arithmetic.

So Peano's axioms are incomplete (assuming that they are consistent). However, the method is much more flexible. It can be used to show that any formal system in which the successor, addition, and multiplication of the natural numbers are represented will be subject to the same incompleteness. Only the details of the Gödel numbering have to be changed, and the proof of the technical Theorem 5.7 has to be re-worked.

In particular, let us adjoin the sentence ζ of Gödel's proof as a new axiom for arithmetic. If the old axioms were consistent, then so are the new ones, since ζ is true in $\mathbb{N}$. But re-doing the proof of Gödel's Theorem, we obtain a new sentence ζ' which is true but not provable from the new axioms.

It might be objected that there is an obvious set of axioms for the natural numbers, which does not suffer from incompleteness: namely, $\mathrm{Th}(\mathbb{N})$, the set of all sentences true in $\mathbb{N}$. Every true sentence is provable in one line from $\mathrm{Th}(\mathbb{N})$. How is the paradox resolved? One of our requirements for a formal system is that the axioms should be mechanically recognizable. This property is crucial in the proof of Gödel's Theorem, though its use is hidden away in the details which we omitted (the construction of a formula which 'recognizes' axioms). So we conclude that the true sentences about the natural numbers cannot be recognized by any mechanical method. Perhaps we should take comfort from this: mathematicians will not be replaced by computers.

The sentence ζ of Gödel's proof is of great importance in the 'metatheory' of the natural numbers; but, if written out as a statement of arithmetic, it would not seem to have any intrinsic interest at all. In the 1970s, Jeff Paris and Leo Harrington discovered a statement of mathematical significance which is true in $\mathbb{N}$ but unprovable in Peano arithmetic. It is a variant on Ramsey's Theorem, which we now describe.

Theorem 5.10 (Finite Ramsey Theorem)

Let k, l, r be positive integers. Then there exists a positive integer n with the property that, if the k-element subsets of $n = \{0, \ldots, n-1\}$ are coloured with r colours, then there is a *monochromatic* subset of size l, that is, a subset all of whose k-element subsets have the same colour.

It is not immediately clear that this theorem can even be formulated in Peano arithmetic, since it requires quantification over subsets and over colourings (which are sets of subsets). We do so as follows. There is a bijection between the set of k-subsets of n and the set $N = \binom{n}{k}$. (This bijection is given by simple arithmetic functions: see Exercise 1.18.) Now a colouring of the elements of N with r colours can be represented by an integer in the range $\{0 \ldots r^N - 1\}$, that is, a member of r^N. (Given x in this range, we express x to base r, obtaining a string of length N of digits in the set $\{0, \ldots, r-1\}$. Now we assign colour j to the point i if and only if the ith digit of x is j.) In this way, Theorem 5.10 can be expressed as a sentence of first-order arithmetic. Moreover, it can be proved from Peano's axioms.

A finite subset X of $\mathbb{N}$ is called *large* if $|X| > \min(X)$. Now the *Paris–Harrington Theorem* reads as follows:

Theorem 5.11 (Paris–Harrington Theorem)

Let k, l, r be positive integers. Then there exists a positive integer n with the property that, if the k-element subsets of $n = \{0, \ldots, n-1\}$ are coloured with r colours, then there is a large monochromatic subset of size at least l.

This result is obviously stronger than the finite Ramsey Theorem. A large set may be hard to find: the further we have to go for the first element of the set, the larger the set must finally be. Paris and Harrington showed that the theorem is not provable in Peano arithmetic.

It is true, however, in the sense that it is a mathematical theorem. This follows from its relationship with the infinite version of Ramsey's Theorem.

Theorem 5.12 (Infinite Ramsey Theorem)

Let k, r be positive integers. Then, if the k-element subsets of $\mathbb{N}$ are coloured with r colours, then there is an infinite monochromatic subset of $\mathbb{N}$.

A proof of this theorem is outlined in Exercise 5.5. In fact the proof can be given in the formal system ZFC for set theory described in the next chapter. Now the finite Ramsey Theorem can be deduced from the infinite using the Compactness Theorem, in much the same way as we proved the infinite Four-Colour Theorem from the finite. (Assume that the finite Ramsey Theorem is false. Then there is some fixed l so that there exist colourings of the k-subsets of arbitrarily large finite sets having no monochromatic l-set. Now the Compactness Theorem shows that there is a colouring of the k-subsets of an infinite set with the same property, contrary to the infinite Ramsey Theorem.) With a little more care, the Paris–Harrington Theorem can also be deduced from the infinite Ramsey Theorem.

We conclude that the Paris–Harrington Theorem is a sentence of first-order arithmetic which is true in $\mathbb{N}$, unprovable in Peano arithmetic, but provable in Zermelo–Fraenkel set theory: a 'natural incompleteness in Peano arithmetic', as the title of the paper of Paris and Harrington in the *Handbook of Mathematical*

Logic [3] has it. For further discussion of the fascinating aspects of Ramsey's
Theorem, see Graham *et al.* [16].

5.4 Consistency

> ... in one sense, a foundation is a security blanket: if you meticu-
> lously follow the rules laid down, no paradoxes or contradictions will
> arise. In reality there is now no guarantee of this sort of security ...

Saunders MacLane, Mathematics: Form and Function [36]

Hilbert hoped to remove contradiction from the foundations of mathematics
by finding a formal proof of its consistency. Such a proof would involve only
finite, mechanical reasoning. The reason for this was Hilbert's view that a
mathematical object exists provided that it is not self-contradictory. Thus, in
Hilbert's example, a real number whose square is −1 does not exist, since these
properties are self-contradictory; Euclidean geometry exists, because Euclid's
axioms are not self-contradictory. The proof of the last assertion is that Euclid's
axioms can be satisfied in the set of real numbers; so really the assertion is that
Euclidean geometry exists if the real numbers do. Since we can represent real
numbers by Dedekind cuts of rational numbers, and so on until we arrive at
the natural numbers, Hilbert regarded it as important to establish consistent
axioms for the natural numbers.

This hope, also, was destroyed by Gödel.

A first-order theory Σ is consistent if no contradiction can be proved from Σ.
The negative nature of this definition reflects the fact that, while it is possible
to demonstrate formally that Σ is inconsistent (by proving a contradiction from
it), there is in general no constructive way to prove in the formal system that Σ
is consistent. We have to step outside the formal system and use the Soundness
and Completeness Theorem: Σ is consistent if we can construct a model for it.

However, the procedure of Gödel numbering opens the possibility of a proof
of consistency within the system. Take any contradiction, for example $0 = 1$,
and let N be its Gödel number (so that $N = 40674055$ in base 12). Then, if π
is the formula describing provability in Theorem 5.7, the sentence

$$(\forall x)(\neg \pi(x, N))$$

asserts the consistency of the system.

The formalists hoped that the consistency of such systems as Peano arithmetic and Zermelo–Fraenkel set theory could be proved by formal methods.
Gödel's Second Incompleteness Theorem shows that this cannot be done.

Theorem 5.13 (Gödel's Second Incompleteness Theorem)

If Peano arithmetic (PA) is consistent, then the formula asserting its consistency is unprovable.

Proof

Let γ denote this formula, and ζ the undecidable formula from Gödel's incompleteness theorem. We proved that, if PA is consistent, then ζ is not provable. Since ζ asserts its own unprovability, this means that ζ is true. Formalising this argument actually gives

$$PA \cup \{\gamma\} \vdash \zeta,$$

since γ asserts the consistency of PA. By the Deduction Theorem,

$$PA \vdash (\gamma \to \zeta).$$

Now, if γ were provable from PA, then we would have a proof of ζ, by one more application of *Modus Ponens*. This is a contradiction; so γ is not provable. □

This means that we cannot be sure that a proof that $0 = 1$ in PA will never be found. (In fact, working mathematicians prove $0 = 1$ every other day; but so far, at least, they have always gone back and checked their working, and found a mistake.)

As in the last section, this result is not specific to Peano arithmetic: any consistent formal system in which the theory of successor, addition and multiplication in $\mathbb{N}$ can be formulated is incapable of proving its own consistency.

On the other hand, we have seen that the natural numbers can be constructed (and the Peano axioms verified) in set theory. So the Zermelo–Fraenkel axioms for set theory (which we study in the next chapter) are strong enough to prove the consistency of Peano's axioms (suitably translated).

EXERCISES

5.1   Use the Compactness Theorem to prove that a graph with well-ordered vertex set can be vertex-coloured with r colours if and only if every finite subgraph can be edge-coloured with r colours, for any positive integer r. (A vertex-colouring is required to have the property that adjacent vertices get different colours.)

5.2   Show that any model of the axioms (P1) and (P2) for the successor function has the form $\mathbb{N} \cup (\mathbb{Z} \times X)$ for some set X, where

0 is the zero element of $\mathbb{N}$, and the successor function is given by

$$s(n) = n + 1 \qquad \text{for } n \in \mathbb{N},$$
$$s((m, x)) = (m + 1, x) \qquad \text{for } (m, x) \in (\mathbb{Z} \times X).$$

5.3 (a) Write the formula $2 + 2 = 4$ in the first-order language of arithmetic, and calculate its Gödel number.

(b) Give a proof of this formula from Peano's axioms. Estimate the Gödel number of your proof, and compare it with the estimated number of elementary particles in the observable universe (about 10^{80}).

5.4 Let M be a model of Peano's axioms, containing an element c which is greater than every natural number. Prove that, if M is represented as $\mathbb{N} \cup (\mathbb{Z} \times X)$ as in Exercise 5.2, the set X indexing copies of $\mathbb{Z}$ is infinite.

5.5 Prove the infinite Ramsey Theorem.

Hint: We use induction on k. For $k = 1$, the assertion is simply the *Pigeonhole Principle*: if the elements of $\mathbb{N}$ are placed in a finite number of pigeonholes, then some pigeonhole contains infinitely many numbers. So assume that the result is true with $k - 1$ replacing k, and let the k-subsets of $\mathbb{N}$ be coloured with finitely many colours $c_0, \ldots, c_{r-1}$.

- Construct an infinite increasing sequence $x_0, x_1, \ldots$ of natural numbers, an infinite decreasing sequence $X_0, X_1, \ldots$ of subsets of $\mathbb{N}$, and an infinite sequence $c_0, c_1, \ldots$ of colours such that $x_i = \min(X_i)$ and any k-subset of X_i containing x_i has colour c_i.

- Use the Pigeonhole Principle to choose a constant subsequence of $(c_0, c_1, \ldots)$. The corresponding subsequence of $(x_0, x_1, \ldots)$ is the required monochromatic set.

5.6 Use the Compactness Theorem to deduce the finite Ramsey Theorem (and the Paris–Harrington Theorem) from the infinite Ramsey Theorem, using the method suggested in the text.

5.7 (a) Show that, for $k = 2$, $r = 2$, $l = 3$, the conclusion of the finite Ramsey Theorem holds if we take $n = 6$. (This means simply that, if we take six points and colour the edges joining them with two colours in any manner, then we must create a monochromatic triangle.)

(b) Show that, for $k = 2$, $r = 2$, $l = 3$, the conclusion of the Paris–Harrington Theorem holds if we take $n = 6$. (The only 3-subset of $\{0, \ldots, 5\}$ which is not large is $\{3, 4, 5\}$; so it is enough to prove that, in any colouring of the edges with two colours, there are at least two monochromatic triangles.)

6
Axiomatic set theory

A choice of axioms is not purely a subjective task. It is usually expected to achieve some definite aim – some specific theorem or theorems are to be derivable from the axioms – and to this extent the problem is exact and objective. But beyond this there are always other important desiderata of a less exact nature: the axioms should not be too numerous, their system is to be as simple and transparent as possible, and each axiom should have an immediate intuitive meaning by which its appropriateness can be judged directly.

John von Neumann and Oskar Morgenstern, *Theory of Games and Economic Behavior* [38]

In this chapter, we present the Zermelo–Fraenkel axioms for set theory, and sketch the justification of them from the Zermelo hierarchy of Chapter 2. The axiom whose status is least clear is the Axiom of Choice. As a result, it has received special attention from mathematicians, and consequences of its truth or falsity have been noted in various parts of mathematics. We will consider some of these. We also develop a theory of infinite cardinal numbers, based on the Axiom of Choice, and say a few words about other systems of axioms which have been proposed.

6.1 Axioms for set theory

As we saw in Chapter 1, everything about sets can be expressed in terms of the membership relation. So the first-order language in which we write the axioms of set theory contains a single binary relation. As usual, we write $x \in y$ instead of $(x, y) \in R$.

The ten Zermelo–Fraenkel axioms are as follows.

1. (*Extension Axiom*) If two sets have the same members, they are equal.

2. (*Empty Set Axiom*) There exists a set $\emptyset$ with no members.

3. (*Pair Set Axiom*) If x and y are sets, then there is a set $\{x, y\}$ whose only members are x and y.

4. (*Union Axiom*) If x is a set, there is a set $\bigcup x$ whose members are the members of members of x.

5. (*Power Set Axiom*) If x is a set, there is a set $\mathcal{P} x$ whose members are the subsets of x.

6. (*Axiom of Infinity*) There is a set a such that $\emptyset \in a$ and $(x \in a) \Rightarrow (\{x\} \in a)$.

7. (*Selection Axiom*) Let ϕ be a first-order formula in the language of set theory with one free variable x, and let a be a set. Then there exists a set b consisting of all members of a which satisfy $\phi(x)$: that is, $b = \{x \in a : \phi(x)\}$.

8. (*Replacement Axiom*) Let ψ be a first-order formula with two free variables x and y which 'defines a partial function': that is, for all x, there is at most one y which satisfies the formula. Let a be any set. Then there is a set b consisting of all those y such that $\psi(x, y)$ holds for some $x \in a$: that is, $b = \{f(x) : x \in a\}$, where f is the 'function defined by ψ'.

9. (*Foundation Axiom*) For any non-empty set x, there exists $y \in x$ with $x \cap y = \emptyset$.

10. (*Axiom of Choice*) If $F : X \to Y$ is a function such that $F(x) \neq \emptyset$ for all $x \in X$, then there exists a function $f : X \to \bigcup Y$ such that $f(x) \in F(x)$ for all $x \in X$.

The statements above of the axioms are informal, but each can be translated into sentences of the first-order language. For example, the Extension Axiom reads

$$(\forall x)(\forall y)((x = y) \leftrightarrow (\forall z)((z \in x) \leftrightarrow (z \in y)))$$

while the Union Axiom reads

$$(\forall x)(\exists y)(\forall z)((z \in y) \leftrightarrow (\exists w)((w \in x) \wedge (z \in w))).$$

In the case of axioms referring to functions, the translation is not so straight-forward. The Pair Set Axiom, applied thrice, guarantees that the *ordered pair* $(x, y) = \{\{x\}, \{x, y\}\}$ exists. The statement 'x is an ordered pair' can then be expressed as a first-order formula. Now the Cartesian product $X \times Y$ can be found as a subset of $\mathcal{P}\mathcal{P}\bigcup\{X, Y\}$ by means of the Selection Axiom. Then the statement 'f is a function from X to Y' can be expressed as a first-order formula. This formula can then be used in the formal statement of the Axiom of Choice. (This is not required for the Axiom of Replacement, since it ap-plies only to functions which are themselves defined by formulae. Thus, only instances involving a set a and formula ϕ such that

$$(\forall x)((x \in a) \rightarrow (\exists y)\phi(x, y))$$

and

$$(\forall x)(\forall y)(\forall z)((\phi(x, y) \wedge \phi(x, z)) \rightarrow (y = z))$$

are considered.)

Probably you feel that the informal statements are clearer!

Remarks

0. We denote the set of ten axioms by ZFC, and the first nine (not including the Axiom of Choice) by ZF.

1. The Extension Axiom implies that the sets defined in the Empty Set, Union, Power Set, Selection and Replacement Axioms are unique. This justifies our previous notation for empty set, power set and union. (Strictly, if such a notation occurs in a formula, it should be replaced by a first-order 'defi-nition' based on the appropriate axiom.) For example, we could translate $\varnothing \in a$ in the Axiom of Infinity by $(\forall x)((\forall y)(\neg(y \in x)) \rightarrow (x \in a))$.

2. The Empty Set Axiom can be weakened to the statement 'There is a set'. For, if a is any set, then $\{x \in a : \neg(x = x)\}$ is a set (by the Selection Axiom), and is empty. In similar fashion, the Axiom of Infinity could be stated informally as 'There exists an infinite set'; but this cannot be written as a first-order sentence. Note that the Axioms of Infinity and Selection imply the Empty Set Axiom. So the axioms are not all independent.

3. The Pair Set axiom follows from the Replacement Axiom and the existence of a set with two elements (which can be taken to be $\mathcal{P}\mathcal{P}\varnothing$): given any two sets x and y, there is a formula in two variables which is true just for the pairs $(\varnothing, x)$ and $(\{\varnothing\}, y)$. Selection follows from Replacement. For, given ϕ, let ψ be the formula

$$(\phi(x) \wedge (y = x)).$$

The (partial) function f defined by ψ maps elements satisfying ϕ to themselves, so
$$\{f(x) : x \in a\} = \{x \in a : \phi(x)\}.$$

Replacement tells us that, if a logical construction does not increase size, then it produces new sets from old.

4. The Selection and Replacement Axioms are not single axioms of first-order logic (unlike the others). Each is an infinite 'axiom scheme' containing one axiom for each choice of the formula ϕ.

5. The Axiom of Foundation is equivalent to the statement that there are no infinite descending chains
$$\ldots \in x_2 \in x_1 \in x_0$$

under the membership relation. This is a metatheorem, since the non-existence of such descending chains cannot itself be stated by a first-order formula. The proof is as follows.

Suppose first that an infinite descending chain of the type shown exists. Take $x = \{x_n : n \in \mathbb{N}\}$. Let y be any element of x. Then $y = x_n$ for some n, and so $x_{n+1} \in y \cap x_n$, contradicting Foundation.

Conversely, suppose that the Axiom of Foundation is false, and let x be a set which is a counterexample: that is, $x \neq \varnothing$, but for all $y \in x$ we have $x \cap y \neq \varnothing$. Now take $x_0 = x$, and define inductively x_{n+1} to be an element of the non-empty set $x \cap x_n$. These elements form the required infinite descending chain. (By induction we show that $x_n \in x$ for all x, so that $x \cap x_n \neq \varnothing$ by assumption.)

The non-existence of infinite descending chains is a simple consequence of Zermelo's hierarchy discussed in Chapter 2. For suppose that such a chain
$$\ldots \in x_2 \in x_1 \in x_0$$

exists, and that V_α is the first stage of the hierarchy containing the top element x_0. As we noted, α is a successor ordinal. Moreover, we can suppose that α has been chosen as small as possible; that is, no infinite descending chain has its top element in an earlier stage than V_α. But now, if $\alpha = s(\beta)$, the element x_1 is in V_β and is the top element of an infinite descending chain obtained just by removing x_0, a contradiction.

6. All the axioms except Choice can be justified informally from the Zermelo hierarchy. The Axiom of Extension really just says that everything is a set, as we remarked earlier. The Empty Set and Infinity Axioms assert the

existence of specific sets (which arise at stages 1 and $\omega+1$ respectively). The Pair Set, Union, Power Set and Replacement Axioms give us new sets from old ones. The Power Set Axiom is implicit in the way that the hierarchy is defined. Consider, for example, the union axiom. If x first appears at stage α, then all its elements have appeared by stage β (where β is the predecessor of α), and all their members by some stage earlier than β. So $\bigcup x$ will appear at stage β or earlier.

The Selection Axiom also gives us new sets, but we can think of its role a bit differently: it helps to clarify the concept of the power set of a set (which is fundamental in the Zermelo hierarchy) by guaranteeing that any subcollection which can be 'described' in first-order language will actually form a subset. (As we said earlier, the vagueness in Zermelo's approach lies in the question: *What is a subset of a set?*)

Finally, the Axiom of Foundation guarantees that every set arises at some point in the hierarchy. For if a set x_0 does not arise, then some member x_1 of x_0 also does not arise (else x_0 would arise at the stage immediately after that when all its elements occur); then some member x_2 of x_1 would not arise; and so on, leading to an infinite descending chain under the membership relation, which is precisely what Foundation forbids.

7. Note that the Axiom of Choice is actually used in showing that failure of the Axiom of Foundation leads to an infinite descending chain. (We have to choose the infinitely many elements $x_0, x_1, x_2, \ldots$). Does this matter?

We accept the axioms because of their justification from Zermelo's hierarchy. Moreover, experience has shown that we have captured enough of Zermelo's construction to form a basis for mathematics: for example, the results of Chapter 1 can all be stated and proved in ZFC.

Once we have justified the axioms, we do not need the Zermelo hierarchy any more; we can feel secure that Russell's Paradox has been avoided. Let us note immediately that in fact no set can be a member of itself. Suppose that y is any set whatsoever. Let $x = \{y\}$ (this exists by the Pair Set Axiom). The only member of x is y, so according to the Axiom of Foundation, $x \cap y = \varnothing$. But $y \in x$; so we must have $y \notin y$, as required.

Hence Russell's set, if it exists, would be the 'set of all sets'. However, if there were a set S consisting of all sets, then S would certainly include itself, contrary to what we have just proved.

6.2 The Axiom of Choice

As mentioned earlier, the Axiom of Choice is the most controversial of the ten Zermelo–Fraenkel axioms. However, large areas of mathematics depend on using it. We will describe two equivalent statements to the Axiom of Choice (abbreviated AC in what follows), and several applications of it in areas such as algebra and measure theory.

The *Well-Ordering Principle* (WO) asserts:

> Every set can be well-ordered.

A *chain* in a partially ordered set $(X, <)$ is a subset Y of X such that Y is totally ordered by the restriction of the relation $<$ (that is, any two members y, z of Y are *comparable*, in the sense that either $y \leq z$ or $z \leq y$). An *upper bound* for a subset C is an element x such that $c \leq x$ for all $c \in C$; we don't require that $x \in C$.

Zorn's Lemma (ZL) asserts:

> Let $(X, <)$ be a partially ordered set in which every chain has an upper bound. Then X has a maximal element.

(Note that the empty set is a chain, and the assumption that it has an upper bound is just the assumption that X is non-empty.)

Theorem 6.1

In ZF, the following are equivalent:

(AC) the Axiom of Choice;

(WO) the Well-Ordering Principle;

(ZL) Zorn's Lemma.

Proof

WO implies AC: Suppose that WO holds, and $F : X \to Y$ is a function with $F(x) \neq \varnothing$ for all $x \in X$. Well-order $\bigcup Y$, and let $f(x)$ be the least element of $F(x)$. Then f is a choice function for F. (The point is that, in a well-ordered set, we have a rule for choosing elements, as with Bertrand Russell's shoes: we simply choose the least element of every non-empty subset that we have to consider!)

AC implies ZL: Let $(X, <)$ be a non-empty partially ordered set in which every chain has an upper bound, and suppose that X has no maximal element. Note that $X \neq \varnothing$, since by assumption the empty chain has an upper bound (in X). Let f be a choice function for the family of non-empty subsets of X. We again attempt to define a map H from ordinals to X. We set

- $H(0) = f(X)$;

- $H(s(\alpha)) = f(\{x \in X : x > H(\alpha)\})$;

- $H(\lambda) = f(Y)$, where Y is the set of all upper bounds for the chain $H[\lambda]$.

We already noted that $X \neq \varnothing$, so the first clause works. The second works because, if $\{x \in X : x > H(\alpha)\}$ were empty, then $H(\alpha)$ would be a maximal element of X, contrary to assumption. The last clause works by assumption once we observe that $H[\lambda]$ really is a chain: in fact we have $H(\alpha) < H(\beta)$ for $\alpha < \beta$.

So there is an injective function from the set of all ordinals to X, which is impossible.

ZL implies WO: Assume that Zorn's Lemma holds. Let X be a set; we wish to show that it can be well-ordered (or, what amounts to the same thing, to find a bijection between X and an ordinal). Let $\mathcal{F}$ be the set of ordered pairs (X', f'), where X' is a subset of X and f' is a bijection from X' to an ordinal. We define a relation $\leq$ on $\mathcal{F}$ as follows: if $(X', f'), (X'', f'') \in \mathcal{F}$, then set $(X', f') \leq (X'', f'')$ if and only if

- $X' \subseteq X''$;

- $f'(x) = f''(x)$ for all $x \in X'$.

If we regard functions as sets of ordered pairs (as indeed they are!), then this says that $f' \subseteq f''$. Then it is easy to see that the relation $\leq$ is a partial order on $\mathcal{F}$.

We claim that the hypotheses of Zorn's Lemma hold, that is, that every chain has an upper bound. So let $\mathcal{C}$ be a chain. Let f be its union (regarding each function, as before, as a set of ordered pairs). We claim that f is a function from a subset of X to an ordinal. It could only fail to be a function if, for some x, we had both (x, y') and (x, y'') in f. This would mean that two functions f' and f'' of $\mathcal{C}$ would satisfy $f'(x) = y'$ and $f''(x) = y''$. But $\mathcal{C}$ is a chain, so one of f' and f'' is smaller than the other, whence $f'(x) = f''(x)$ by the definition of the order. Now the range of f is the union of the ranges of the members of $\mathcal{C}$, each of which is an ordinal, so is an ordinal; and thus (X, f) is an element of $\mathcal{F}$, and is an upper bound for the chain, as required.

By Zorn's Lemma (our assumption), there is a maximal element (X_0, f_0). We claim that $X_0 = X$. If not, choose $x \in X \setminus X_0$, let α be the range of f_0, and define $X_1 = X_0 \cup \{x\}$ and $f_1 = (x, \alpha)$ – in other words, we extend f_0 to a larger function which maps x to α. Then $f_0 < f_1$, contradicting the maximality of f_0. So f is a bijection from X to an ordinal, as required. $\square$

Now we turn to some of the principal applications of the Axiom of Choice in mathematics. I will give three applications, the methods of proof illustrating the use of the three equivalent principles of Theorem 6.1.

A *maximal ideal* of a ring R is an ideal I with the property that, for any ideal J containing I, either $J = I$ or $J = R$. See Wallace [46], Chapter 5.

Theorem 6.2

In ZFC, every ring with identity has a maximal ideal.

Proof

Consider the set $\mathcal{I}$ of proper ideals of R (that is, ideals not equal to R itself), ordered by inclusion. This is a partially ordered set. We verify the hypotheses of Zorn's Lemma. So let $\mathcal{C}$ be a chain in $\mathcal{I}$. If $\mathcal{C} = \varnothing$, then $\{0\}$ is a proper ideal and is an upper bound for $\mathcal{C}$. So suppose that $\mathcal{C} \neq \varnothing$, and let $I = \bigcup \mathcal{C}$. We claim that $I \in \mathcal{I}$: if so, then I is clearly an upper bound for $\mathcal{C}$.

- *I is closed under addition*: Take $x_1, x_2 \in I$. Then $x_1 \in I_1$ and $x_2 \in I_2$ for some $I_1, I_2 \in \mathcal{C}$. Since $\mathcal{C}$ is a chain, we can assume without loss of generality that $I_1 \subseteq I_2$. Then $x_1, x_2 \in I_2$, and so $x_1 + x_2 \in I_2$ (since I_2 is an ideal), whence $x_1 + x_2 \in I$.

- *$x \in I$, $r \in R$ imply $xr, rx \in I$*: Similar to the above but easier.

- *I is a proper ideal*: We use the fact that an ideal containing the identity element 1 is the whole ring. Since all the ideals in $\mathcal{C}$ are proper, none of them contains 1, and so their union I doesn't contain 1.

By Zorn's Lemma, $\mathcal{I}$ contains a maximal element, which is a maximal ideal of R. $\square$

A *basis* in a vector space V is a subset B of V with the property that every element of V has a unique expression as a linear combination of a *finite* subset of B. (We have to put finiteness into the definition since B might be infinite. By 'unique expression' we disallow adding extra vectors with zero coefficient; that is, we require that all coefficients in the linear combination are non-zero.) An equivalent definition is that B *spans* V (that is, every vector is a linear combination of a finite subset of B), and that B is *linearly independent*, that is, if a linear combination of a finite subset of B is zero, then all the coefficients are zero. See Blyth and Robertson [7] for more explanation.

Theorem 6.3

In ZFC, every vector space has a basis.

Proof

Well-order V: that is, write it as $V = \{v_\beta : \beta < \alpha\}$ for some ordinal α. Now we construct B_β for $\beta \leq \alpha$ as follows:

- $B_0 = \varnothing$.

- If v_β is a linear combination of a finite number of vectors from B_β, then we put $B_{s(\beta)} = B_\beta$; if not, we put $B_{s(\beta)} = B_\beta \cup \{v_\beta\}$.

- If $\lambda \leq \alpha$ is a limit ordinal, then

$$B_\lambda = \bigcup_{\beta < \lambda} B_\beta.$$

We claim that $B = B_\alpha$ is a basis for V.

- B *spans* V: Every vector of V is of the form v_β for some $\beta < \alpha$. Either v_β is a linear combination of the vectors in B_β, or $v_\beta \in B_{s(\beta)}$; in either case, v_β is a linear combination of vectors in B.

- B *is linearly independent*: Suppose that $\sum c_\beta v_\beta = 0$ where the sum involves a finite number of elements of B with non-zero coefficients. Let v_β be the last vector in the ordering which has a non-zero coefficient. Then from this equation we can express v_β as a linear combination of earlier vectors. But according to the definition, this means that $v_\beta \notin B$, a contradiction.   □

The last example concerns *Lebesgue measure* μ, a way to 'measure' the size of a set of real numbers. This is probably less familiar. All we require is that

- the measure of a bounded set of real numbers (if it exists) is a non-negative real number;

- if $A \subseteq B$ then $\mu(A) \leq \mu(B)$ (if the measures exist);

- the measure of an interval is its length;

- the measure is *countably additive*, in the sense that

$$\mu\left(\bigcup_{n \in \mathbb{N}} X_n\right) = \sum_{n \in \mathbb{N}} \mu(X_n)$$

if the sets X_n are pairwise disjoint; and

• the measure of a set is unaffected by translation.

Now the measure of a singleton set $\{x\}$ is zero (since this set is a closed interval $[x, x]$ of length zero). It follows by countable additivity that the measure of a countable set is zero.

Theorem 6.4

In ZFC, there exist non-measurable bounded sets of real numbers.

Proof

Define a relation on the unit interval $[0, 1]$ by the rule that $x \sim y$ if $y - x$ is a rational number. This is easily seen to be an equivalence relation. Let C be the set of equivalence classes of this relation (that is, the corresponding partition of $[0, 1]$). By AC, there is a set S containing one element from each equivalence class. (We apply AC to the identity function $F : C \to C$, to obtain a choice function $f : C \to \bigcup C = [0, 1]$ with $f(c) \in C$ for each class C; then take $S = f[C] = \{f(c) : c \in C\}$. See Exercise 1.19.)

The translates $S_q = S + q$, for $q \in [-1, 1] \cap \mathbb{Q}$, are pairwise disjoint. For, if $x \in (S + q_1) \cap (S + q_2)$, then $x = s_1 + q_1 = s_2 + q_2$ with $s_1, s_2 \in S$; but then $s_1 - s_2 = q_2 - q_1 \in \mathbb{Q}$, contradicting the fact that S is a set of distinct representatives for the equivalence classes. Now each set S_q is contained in the interval $[-1, 2]$; and

$$[0, 1] \subseteq S^* = \bigcup_{q \in [-1,1] \cap \mathbb{Q}} S_q \subseteq [-1, 2],$$

since any number x in $[0, 1]$ is equivalent to a unique element $s \in S$, and so $x - s$ is rational, with $-1 \le x - s \le 1$, since both x and s lie in $[0, 1]$.

Now the assumption that S is measurable leads to a contradiction. For suppose that $\mu(S) = c$. Then, by translation-invariance, $\mu(S_q) = c$ for all $q \in [-1, 1] \cap \mathbb{Q}$. If $c = 0$ then, by countable additivity, $\mu(S^*) = 0$, contradicting the fact that this S^* contains $[0, 1]$, with measure 1. On the other hand, if $c > 0$, then $\mu(S^*) = \infty$, contradicting the fact that S^* is contained in $[-1, 2]$, with measure 3.

We conclude that S is a non-measurable set.   □

Based on the existence of non-measurable sets, Banach and Tarski demonstrated the *Banach–Tarski Paradox*: it is possible to partition a unit sphere into a finite number of pieces, which can be re-assembled (by rigid motions) to form two unit spheres, or a sphere of radius 2. For the proof, see Wagon [45].

You may feel that this unintuitive result shows that the Axiom of Choice is unacceptable. Of course, the pieces in the decomposition must be non-measurable sets.

We saw in Chapter 3 that the Compactness Theorem holds for propositional logic with a finite or countable set of propositional variables, and remarked that the proof goes through for any well-ordered set of propositional variables. Hence:

Theorem 6.5

In ZFC, the Compactness Theorem holds for propositional logic based on any set of propositional variables.

We give two different proofs of this result.

First Proof

In ZFC, the set of propositional variables can be well-ordered; we saw that the Soundness and Completeness Theorem then holds for the logic, and the Compactness Theorem follows from this in the usual way.   □

Second Proof

This argument is more algebraic: see Wallace [46], Chapter 5, for the ring-theoretic background.

Let P denote the set of propositional variables, and Σ a set of formulae with the property that every finite subset of Σ is consistent. As in Section 3.4, consider the Boolean algebra $B(P)$ of equivalence classes of propositional formulae in the variables P. By Theorem 3.11, we can re-interpret $B(P)$ as a Boolean ring, so that valuations are ring homomorphisms from $B(P)$ to $R_2 = \mathbb{Z}/2\mathbb{Z}$. We can regard Σ as a set of elements of the ring $B(P)$. Let I be the ideal generated by $\{1 + s : s \in \Sigma\}$. We separate the argument into two cases.

Case 1: $I = B(P)$. Then the element 1 lies in I, so we have an equation

$$1 = r_1(1 + s_1) + \cdots + r_n(1 + s_n),$$

where $r_1, \ldots, r_n \in B(P)$ and $s_1, \ldots, s_n \in \Sigma$. Now $\{s_1, \ldots, s_n\}$ is consistent, by hypothesis; so there is a homomorphism $v : B(P) \to R_2$ such that $v(s_i) = 1$ for $i = 1, \ldots, n$. But then applying the homomorphism to the displayed equation, we find that $1 = 0$ (in R_2), a contradiction. So this case cannot arise.

Case 2: $I \subset R(P)$. By Theorem 6.2 (which uses the Axiom of Choice), there is a maximal ideal J of $B(P)$ containing I. Since $B(P)$ is a commutative ring with identity, $B(P)/J$ is a field, all of whose elements satisfy the polynomial $x^2 - x$ of degree 2. So $|B(P)/J| = 2$, and $B(P)/J \cong R_2$. So there is a homomorphism $v : B(P) \to R_2$ whose kernel is J. Thus, there is a valuation v such that $v(1 + s) = 0$ (in other words, $v(s) = 1$) for all $s \in \Sigma$; that is, Σ is satisfiable, as required.   □

Remark We have seen that various mathematical facts (such as the infinite Four-Colour Theorem, and the fact that every set can be totally ordered) can be proved using the Propositional Compactness Theorem. So these are also consequences of the Axiom of Choice. It is known, however, that Propositional Compactness is a 'weaker' principle than AC: there is no proof of AC using Propositional Compactness, and indeed models of set theory have been constructed in which Propositional Compactness is true but AC fails. See the chapter by John Truss in Kaye and Macpherson [29] for a survey of this.

6.3 Cardinals

We now develop a theory of cardinal numbers. As in the case of the ordinals, we can state at the start the theorem that we want to prove from our definition. Since cardinal numbers should measure the size of arbitrary sets, we require a theorem which says:

Theorem 6.6

Every set has a bijection to a unique cardinal number.

It has to be said that no really adequate theory of cardinal numbers exists in ZF. Bertrand Russell attempted a definition in which, for example, the number 2 is the class of all 2-element sets. With this definition, however, 2 is not even a set, and certainly not a 2-element set! However, with the Axiom of Choice, things are much simpler. We work in ZFC for the rest of this section. (In ZF, this theory applies to those sets which can be well-ordered.)

Definition 6.1

A *cardinal* is an ordinal α with the property that there is no bijection between α and any section of α.

Note that, according to this definition, all finite ordinals (that is, all natural numbers) are cardinals; and ω is a cardinal, since it is infinite but all its sections are finite. However, $\omega + 1$ is not a cardinal, since it is countable (that is, has a bijection to its section ω).

Proof of Theorem 6.6 (in ZFC)

Let X be a set. By WO, X can be well-ordered; that is, there is a bijection from X to some ordinal. Now there is a smallest ordinal α in the set of ordinals bijective with X. And α is a cardinal; for, if there was a bijection from α to a section β, then there would be a bijection from X to β, contrary to the choice of α.

Now, if X has a bijection to two cardinal numbers α and β, then there is a bijection between α and β, contradicting the fact that the smaller is a section of the larger. $\square$

We denote the cardinal of the set X (the unique cardinal bijective with X) by $|X|$. Note that, if α is a cardinal, then $|\alpha| = \alpha$.

Cantor introduced the *aleph notation* for infinite cardinals. (The letter $\aleph$, 'aleph', is the first letter of the Hebrew alphabet.) This is a function from ordinals to cardinals, defined by transfinite recursion as follows:

- $\aleph_0 = \omega$;

- $\aleph_{s(\alpha)}$ is the smallest cardinal greater than $\aleph_\alpha$;

- if λ is a limit ordinal then

$$\aleph_\lambda = \bigcup_{\beta < \lambda} \aleph_\beta.$$

It is not obvious that $\aleph_\lambda$ is a cardinal. It is certainly an ordinal, since it is a union of ordinals. Suppose that it were bijective with a section of itself. This section could not contain all $\aleph_\beta$; but if some $\aleph_\beta$ does not lie in the section, then the restriction of the bijection takes $\aleph_\beta$ into a section of itself, a contradiction.

So we have two notations for the ordinal describing the infinite sequence of natural numbers, namely ω and $\aleph_0$. We use the first if we are thinking of it as an ordinal, and the second when we regard it as a cardinal. Note that $\aleph_1$ is the first uncountable ordinal.

There is an order relation defined on cardinals, since they are special kinds of ordinals. In fact, we have:

$|X| \le |Y|$ if and only if there is an injective function from X to Y.

For certainly $|X| \leq |Y|$ implies that $|X|$ (as an ordinal) is a subset of $|Y|$. Suppose that an injective function $f : \alpha \to \beta$ exists, where α and β are cardinals. Then there is a bijective function from α to an ordinal not exceeding β (since the image of f is well-ordered and contained in β); so the cardinal α is not greater than β.

The *Schröder–Bernstein Theorem* can be written in terms of cardinals as follows.

Theorem 6.7

For any two sets X and Y, if $|X| \leq |Y|$ and $|Y| \leq |X|$ then $|X| = |Y|$.

Now we will define arithmetic operations (addition, multiplication and exponentiation) of cardinals. The simplest way to do this is to mirror the operations of disjoint union, cartesian product, and set of functions: that is, for cardinals α and β, we define

- $\alpha + \beta = |(\alpha \times \{0\}) \cup (\beta \times \{1\})|$;
- $\alpha \cdot \beta = |\alpha \times \beta|$;
- $\alpha^{\beta} = |\alpha^{\beta}|$;

where in the third (confusing) equation, on the right-hand side A^B means the set of all functions from B to A, and not the ordinal exponentiation defined in Chapter 2. We can write these definitions as statements about the cardinalities of arbitrary sets, as follows:

- $|A \cup B| = |A| + |B|$ if A and B are disjoint;
- $|A \times B| = |A| \cdot |B|$;
- $|A^B| = |A|^{|B|}$.

It turns out that cardinal addition and multiplication tables are very easy to learn!

Theorem 6.8

Let α and β be non-zero cardinals, at least one of which is infinite. Then

$$\alpha + \beta = \alpha \cdot \beta = \max\{\alpha, \beta\}.$$

Proof

We claim that it is enough to prove that

$$\alpha \cdot \alpha = \alpha$$

for any infinite cardinal α. For, if $\beta \leq \alpha$, then there is an obvious injection
from $\alpha \times \beta$ to $\alpha \times \alpha$; and there is an injection from $(\alpha \times \{0\}) \cup (\beta \times \{1\})$ to

$$(\alpha \times \{0\}) \cup (\alpha \times \{1\}) = \alpha \times 2,$$

where $2 = \{0, 1\}$. So $\alpha \cdot \beta \leq \alpha \cdot \alpha$ and $\alpha + \beta \leq \alpha \cdot \alpha$. On the other hand, clearly
$\alpha \leq \alpha + \beta$ and $\alpha \leq \alpha \cdot \beta$ (the latter if $\beta \neq 0$).

We have already proved in Chapter 1 that $\aleph_0 \cdot \aleph_0 = \aleph_0$. The general proof
follows similar lines but is rather more complicated. We suppose that the theo-
rem is false. In that case, there will be a smallest cardinal α such that $\alpha \cdot \alpha > \alpha$.
We let $P = \alpha \times \alpha$. Now recall that α is an ordinal, which means that it is a set
of ordinals. In the following argument, we use *ordinal addition*! We define, for
ordinals $\beta < \alpha$, the subset P_β of P by the rule

$$P_\beta = \{(x, y) \in P : x + y = \beta\}.$$

These sets correspond to the north-east to south-west arrows in Figure 1.3.
We claim that the sets P_β for $\beta < \alpha$ form a partition of P. Clearly they are
pairwise disjoint. To show that every point lies in one of them, we need to
show that, if $x, y < \alpha$, then $x + y < \alpha$: but this follows from the fact that the
theorem is true for cardinals less than α. (The ordinal $x + y$ has cardinality
$|x| + |y| = \max\{|x|, |y|\}$.)

Now we well-order each 'diagonal strip' P_β by the 'lexicographic' rule

$$(x, y) < (x', y') \text{ if either } x < x', \text{ or } x = x', \, y < y'.$$

This is easily seen to be a well-ordering – it is the ordering induced on P_β as a
subset of $\alpha \times \alpha$. So now we can well-order all of P by putting $(x, y) < (x', y')$ if
either $(x, y) \in P_\beta$, $(x', y') \in P_\gamma$ with $\beta < \gamma$, or $(x, y) < (x', y')$ within P_β under
the ordering already defined. This gives a well-ordering of P.

Let θ be the unique ordinal isomorphic to P. We have $\theta > \alpha$, since $|P| > \alpha$
by assumption. So there is a point (u, v) in P such that the section (u, v) is
isomorphic to α. Suppose that $(u, v) \in P_\beta$: that is, $u + v = \beta$. Then all points
$(x, y) \in P$ with $(x, y) < (u, v)$ satisfy $x + y \leq \beta$, whence $x, y \leq \beta$; so this entire
section of P is contained in $s(\beta) \times s(\beta)$. We conclude that

$$|s(\beta) \times s(\beta)| = \alpha > |s(\beta)|,$$

a contradiction since $s(\beta) < \alpha$. (An infinite successor ordinal cannot be a
cardinal, and certainly $|\beta| < \alpha$.)

The theorem is proved. $\square$

The theorem implies, in particular, that the union of at most α sets, each
of cardinality at most α, has cardinality at most α, for any infinite cardinal α.
So, using sums and products, we cannot build ever larger sets.

Exponentiation is much less trivial, however, and certainly has the ability to construct larger sets. We give a brief introduction. First, we have

Theorem 6.9

(a) For any set X, $|\mathcal{P}\,X| = 2^{|X|}$.

(b) $|\mathbb{R}| = 2^{\aleph_0}$.

Proof

(a) We have to produce a bijection between the set of subsets of X and the set of functions from X to $2 = \{0,1\}$. We do this by representing a subset Y of X by its *characteristic function* χ_Y, defined as follows:

$$\chi_Y(x) = \begin{cases} 1 & \text{if } x \in Y, \\ 0 & \text{if } x \notin Y. \end{cases}$$

Then distinct sets have distinct characteristic functions; and any function $F : X \to 2$ is the characteristic function of some set, namely the set $\{x \in X : f(x) = 1\}$. So we have a bijection as required.

(b) First, note that the cardinality of $\mathbb{R}$ is the same as that of the unit interval $(0,1)$: the map $f(x) = \tan \pi(x - \frac{1}{2})$ is a bijection from $(0,1)$ to $\mathbb{R}$. Moreover, we can regard an element of $2^{\mathbb{N}}$ as an infinite sequence of zeros and ones. Now we have an injection from $2^{\mathbb{N}}$ to $(0,1)$ by regarding the infinite sequence as an infinite decimal expansion (which happens to use only zeros and ones); and an injection from $(0,1)$ to $2^{\mathbb{N}}$ by taking the base 2 expansion of a number in the unit interval (resolving ambiguities by assuming that the base 2 expansion of a rational whose denominator is a power of 2 ends with infinitely many zeros rather than with infinitely many ones). By the Schröder–Bernstein Theorem, the cardinalities are equal.   □

Thus, Cantor's Theorem (Theorem 1.10) can be translated into the form

Theorem 6.10

For any cardinal α, $2^\alpha > \alpha$.   □

Cardinal arithmetic satisfies some (but of course not all) of the laws of the arithmetic of the natural numbers. In particular, it is true that $(\alpha^\beta)^\gamma = \alpha^{\beta \cdot \gamma}$, as is shown by producing a bijection between these sets (see Exercise 6.1). This simple observation has the following consequence:

Theorem 6.11

Let α and β be cardinals, with α infinite and $2 \leq \beta \leq 2^\alpha$. Then $\beta^\alpha = 2^\alpha$.

Proof

This follows from

$$2^\alpha \leq \beta^\alpha \leq (2^\alpha)^\alpha = 2^{\alpha \cdot \alpha} = 2^\alpha,$$

on using the Schröder–Bernstein Theorem. □

So much of the mystery of cardinal arithmetic lies in the function $\alpha \to 2^\alpha$. By Cantor's Theorem, we have $2^{\aleph_\alpha} \geq \aleph_{s(\alpha)}$ for any ordinal α. Do we have equality or not? The famous *Continuum Hypothesis* asserts that $2^{\aleph_0} = \aleph_1$. This was one of the problems posed in 1900 to the mathematical community by David Hilbert, to guide the development of mathematics in the twentieth century. In Hilbert's words (as translated by Dr Mary Winston Newson in the *Bulletin of the American Mathematical Society*), quoted in [8],

> Two assemblages, i.e. two assemblages of ordinary real numbers or points, are said to be (according to Cantor) equivalent or of *equal cardinal number*, if they can be brought into a relation to one another such that to every number of the one assemblage corresponds one and only one definite number of the other. The investigations of Cantor on such assemblages of points suggest a very plausible theorem, which nevertheless, in spite of the most strenuous efforts, no one has succeeded in proving. This is the theorem:
>
> Every system of infinitely many real numbers, i.e. every assemblage of numbers (or points), is either equivalent to the assemblage of natural integers, $1, 2, 3, \ldots$, or to the assemblage of all real numbers and therefore to the *continuum*, that is, to the points on a line: *as regards equivalence there are, therefore, only two assemblages of numbers, the countable assemblage and the continuum.*

Here Hilbert is using the term 'assemblage' for our 'infinite set', and considering subsets of $\mathbb{R}$. Since $|\mathbb{R}| = 2^{\aleph_0}$, he asks whether it is true that there is no infinite cardinal strictly between $\aleph_0$ and $2^{\aleph_0}$, that is, whether $2^{\aleph_0} = \aleph_1$. Its plausibility was reinforced when Gödel proved that it cannot be disproved in ZFC. Thirty years later, however, Hilbert was answered in a way he would not have expected by Cohen, who showed that it cannot be proved in ZFC either. By a new technique known as *forcing*, he constructed a model of ZFC in which $2^{\aleph_0} = \aleph_2$.

Much is known about the possible values of $2^{\aleph_\alpha}$. The assertion $2^{\aleph_\alpha} = \aleph_{s(\alpha)}$ is known as the *Generalized Continuum Hypothesis*. It, like the specialised version, is independent of ZFC, and we may assume it true or false as we choose (but the answers to our mathematical questions will depend on the assumption we make).

Finally on the subject of cardinals, here is the statement of the Upward Löwenheim–Skolem Theorem (in ZFC), with a small addition obtained by looking at the proof more carefully:

Theorem 6.12

Let Σ be a set of sentences in a first-order language $\mathcal{L}$. Suppose that Σ has an infinite model. Then, for any cardinal $\alpha \geq |\mathcal{L}|$, Σ has a model with cardinal α.

6.4 Inaccessibility

A cardinal α is said to be *inaccessible* if the following three conditions hold:

(a) $\alpha > \aleph_0$;

(b) for any cardinal $\lambda < \alpha$, we have $2^\lambda < \alpha$;

(c) the union of fewer than α ordinals, each smaller than α, is smaller than α.

Do inaccessible cardinals exist? In a sense, the definition is modelled on $\aleph_0$, and reflects the immense jump from the finite to the infinite: $\aleph_0$ satisfies conditions (b) and (c) (though not, of course, (a)). An inaccessible cardinal would signal a similarly vast jump somewhere in the set-theoretic universe. The cardinal $\aleph_1$ satisfies (a) and (c), since a union of countably many countable sets is countable; but it fails (b), since $2^{\aleph_0} \geq \aleph_1$ by Cantor's Theorem. If we define a sequence $(\beth_\alpha)$ of cardinals by the transfinite recursion

- $\beth_0 = \aleph_0$;

- $\beth_{s(\alpha)} = 2^{\beth_\alpha}$;

- if λ is a limit ordinal then $\beth_\lambda = \bigcup_{\alpha < \lambda} \beth_\alpha$,

then $\beth_\omega$ satisfies (a) and (b), but not (c) (since it is a countable union of smaller cardinals by definition). (Note: $\beth$, 'beth', is the second letter of the Hebrew alphabet. The Generalised Continuum Hypothesis is the assertion that $\aleph_\alpha = \beth_\alpha$ for all ordinals α.)

Inaccessible cardinals derive their importance from the following result. Recall that, for any ordinal α, we let V_α denote the set of sets constructed at stage α of the Zermelo hierarchy.

Theorem 6.13

(a) If α is a limit ordinal, then V_α satisfies all the Zermelo–Fraenkel axioms except possibly the Replacement Axiom.

(b) If α is an inaccessible cardinal, then V_α satisfies all the Zermelo–Fraenkel axioms.

Proof

(a) It is necessary to check that, for those axioms which build new sets out of old, the sets built from sets in V_α themselves lie in V_α. Consider the power set, for example. If x appears at stage β, then $\mathcal{P}\,x$ appears at stage $s(\beta)$; and $\beta < \alpha$ implies $s(\beta) < \alpha$.

(b) Suppose that ϕ is a formula which 'defines a function' in V_α; that is, if $\phi(x, y_1)$ and $\phi(x, y_2)$ hold for $x, y_1, y_2 \in V_\alpha$, then $y_1 = y_2$. Let x be any set of V_α, and let

$$y = \{F(u) : u \in x\},$$

where F is the function defined by ϕ. Now each element $u \in x$ appears at a stage earlier than α, and so the same is true of $F(u)$. Also, the cardinality of x also appears at a stage earlier than α. Now the inaccessibility of α shows that the union of all the stages at which elements $F(u)$ appear for $u \in x$ is smaller than α, so the set y appears before stage α. $\quad\square$

Corollary 6.1

The statement 'an inaccessible cardinal exists' is not provable in ZFC.

Proof

If an inaccessible cardinal α exists, then V_α is a model of ZFC, and so ZFC is consistent. But by Gödel's Second Incompleteness Theorem (Theorem 5.13), the consistency of a theory as powerful as ZFC (which includes the theory of the natural numbers) cannot be proved from the axioms of that theory. $\quad\square$

We remarked that inaccessible cardinals are uncountable generalisations of ω. The corollary above is analogous to the fact that the Infinity Axiom (asserting the existence of ω) cannot be proved from the other axioms of ZFC. Indeed, we see in Exercise 6.11 that a model for the other nine axioms can be built from the natural numbers (themselves constructible in ZFC).

The set V_α, for an inaccessible cardinal α, is an example of a $\in$-*model*, a model of set theory where the membership relation is the restriction of the usual membership relation $\in$ to the domain of the model. It is possible to write down the requirements on a non-empty set V for it to support a $\in$-model. The first requirement is that V is a *transitive set*; that is, if $x \in V$ and $y \in x$, then $y \in V$. In addition, we require closure of V under the constructions specified by the axioms of ZF: that is,

- $\omega \in V$ (this guarantees the Infinity Axiom);

- if $x, y \in V$ then $\{x, y\} \in V$;

- the power set and union of sets in V are in V;

- the image of a member of V under a function defined by a first-order formula is in V.

Part (c) of the definition of inaccessibility can be quantified. The *cofinality* of a cardinal α is the smallest cardinal κ so that α is the union of κ strictly smaller ordinals. More generally, we say that a subset of the ordinal α is *cofinal* in α if its union is α; then the cofinality of α, written $\mathrm{cf}(\alpha)$, is the smallest cardinality of a cofinal set. We see that $\mathrm{cf}(\alpha) \leq \alpha$ for all cardinals α; we call α *regular* if $\mathrm{cf}(\alpha) = \alpha$, and *singular* otherwise.

For example, $\aleph_n$ is regular for all natural numbers n, but $\aleph_\omega$ is singular (indeed, $\mathrm{cf}(\aleph_\omega) = \omega$). These are special cases of the following result:

Theorem 6.14

(a) If α is a successor ordinal, then $\aleph_\alpha$ is regular.

(b) For any limit ordinal λ, $\mathrm{cf}(\aleph_\lambda) = \mathrm{cf}(\lambda)$.

Proof

(a) If $\alpha = s(\beta)$, then any cardinal less than $\aleph_\alpha$ is at most $\aleph_\beta$; and the union of at most $\aleph_\beta$ such cardinals has cardinality at most $\aleph_\beta \cdot \aleph_\beta = \aleph_\beta$.

(b) If S is a set of ordinals with union λ, then

$$\bigcup_{\alpha \in S} \aleph_\alpha = \aleph_\lambda. \qquad \square$$

Note This does not imply that $\aleph_\lambda$ is singular for limit ordinals λ; it may happen that $\aleph_\lambda = \lambda = \mathrm{cf}(\lambda)$. (There is a surprise lurking in this observation.

The cardinals appear, at the start, to grow much faster than the ordinals: compare 0 with $\aleph_0$, 1 with $\aleph_1$. But, at a regular limit cardinal, the tortoise catches up with the hare.)

6.5 Alternative set theories

One of the great intellectual shocks of the nineteenth century (along with Darwinism and the re-evaluation of the age of the Earth by geologists) was the development of non-Euclidean geometry. After two millennia of certainty that Euclid had described the ideal geometry to which our surveying and measuring were approximations, suddenly there was a viable alternative description of space. More worryingly, it could no longer be asserted that the theorems in Euclid's book were *true*. Mathematicians learned to say instead that the theorems were consequences of the axioms, and that different sets of axioms would naturally have different consequences. The intellectual problem (and opportunity) was now one for the natural scientists, who had to decide which geometry best described the universe in which we live. Einstein based his general relativity on a geometry constructed by Riemann.

The discovery of alternative versions of set theory brought the problem much closer to home for the mathematicians. Now they were describing not just alternative models for the real universe, but alternative models for the ideal universe of thought in which mathematicians work. As we have seen, many mathematical results, both natural (every vector space has a basis) and unnatural (there exist non-measurable sets of real numbers) follow from the Axiom of Choice, which cannot itself be justified on the basis of sound intuition about what sets are.

One reaction to this is to investigate the consequences of the controversial axioms and additional principles which have been proposed. For example, the Axiom of Choice is equivalent to the statement that every set can be well-ordered. It implies that any partial order of a set can be 'extended' to a total order (this is proved by applying Zorn's Lemma to the set of all partial orders extending a given one, and showing that a maximal partial order is total). But this principle turns out to be strictly weaker than AC. Weaker still is the principle that every set can be totally ordered (this follows on applying the previous result to the empty partial order on the set). Yet weaker is the principle of *Propositional Compactness*: if Σ is a set of propositional (Boolean) formulae in any number of variables having the property that any finite subset of Σ is satisfiable (that is, the formulae can be simultaneously given the value T by a suitable assignment of values to the variables), then Σ is satisfiable. (We saw that this principle holds for well-ordered sets of propositional variables: this makes it clear that Propositional Compactness is implied by the Well-Ordering Principle.)

We saw that AC implies that every ring with identity has a maximal ideal. Wilfrid Hodges showed the converse: in ZF, the statement that every ring with identity has a maximal ideal implies AC. Weaker principles can be obtained by restricting the classes of rings. For example, the statement that every Boolean ring has a maximal ideal is equivalent to Propositional Compactness.

Various stronger axioms (within ZFC) have been studied, including Martin's Axiom and Jensen's Diamond.

More radical is to delete AC altogether and replace it with a contradictory assertion. It is not easy to find a plausible candidate. One such is the *Axiom of Determinacy* which states that, in any two-player game in which draws are not possible, one player has a winning strategy. If this axiom is added to ZF, we obtain consequences such as the statement that every subset of $\mathbb{R}$ is measurable.

Other axioms have also come in for scrutiny, including the Foundation Axiom: Aczel proposed an alternative *Anti-Foundation Axiom*, which is the subject of the book by Barwise and Moss [4].

Exercise 6.11 gives a model of set theory in which all the axioms of ZFC are valid except for the Infinity Axiom. This is a model for 'finite set theory'.

6.6 The Skolem Paradox

> ... there seems to be a paradox if theories can have countable models and yet there is a theory about uncountable things ... this was very thought-provoking and it led to exceptionally good theorems later on.
>
> John N. Crossley *et al.*, *What is Mathematical Logic?* [10]

In the last section, we considered some alternative axiom systems. But even if we settle on a particular axiom system (such as ZFC), we cannot expect the structure of the set-theoretical universe to be completely determined. We saw that ZFC does not resolve questions such as the Continuum Hypothesis.

One of the most striking examples of this is the *Skolem Paradox*:

Theorem 6.15

There exists a countable model of ZFC.

Indeed this follows trivially from the downward Löwenheim–Skolem Theorem and the facts that the language of set theory is countable (having a single binary relation symbol for membership) and the axioms are first-order. (Strictly speaking, the conclusion is that if ZFC is consistent, then it has a countable model!)

The paradox arises from Cantor's Theorem (Theorem 1.10), which asserts that there exist uncountable sets. How can a countable model contain uncountable sets?

The resolution of this paradox is obtained by looking more closely at what we mean by a model of ZFC. This is just a set, with a binary relation (for 'membership'), in which the ten statements represented by the axioms are all valid. The elements of the model are not themselves sets, but simply elements of the model; the question of their countability or uncountability does not immediately arise. Given an element x, we can think of it as the set of all those elements y (in the model) for which $y \in x$ holds. To help get away from our intuitive thinking about sets, regard the model as a set of points, with an arrow from x to y if the relation $x \in y$ holds. Then x is represented by the set of all elements y for which there is an arrow to x. If the model is countable, this set will certainly be countable!

Cantor's Theorem, however, asserts that there is a set x which is not countable. This means that, *in the model*, there is no bijection between x and the set of natural numbers. To decode this, we have to remember that a function is a set of ordered pairs, and an ordered pair is itself a set. If $x \neq y$, the ordered pair (x, y) is a point for which the configuration shown in Figure 6.1 holds. (There are no incoming arrows at the middle and top levels apart from those shown.) A similar but simpler picture applies if $x = y$.

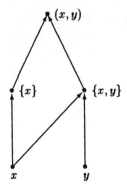

Fig. 6.1. An ordered pair

Now a function is an element of the model for which all the incoming arrows are from points like those at the top of the figure (and where some extra conditions apply). Cantor's Theorem asserts that two particular elements of the model are such that no function of a certain type connects the points for which arrows go to these points. It is a quite complicated combinatorial fact, but no contradiction in mathematics is involved!

6.7 Classes

On several occasions already, we have found it convenient to refer to a collection of sets which is not itself a set. (For example, we called the ordinal numbers a 'well-ordered class' in Chapter 2.) Now we have a language to describe such collections. Russell's Paradox does not mean that it is forbidden to talk about all sets! A collection of sets which is not a set is called a *proper class*.

We can refer to the class of sets having some particular property. This property should be expressible in first-order language in set theory, by means of a formula ϕ with one free variable x. Thus, the class $\mathcal{C}(\phi) = \{x : \phi(x)\}$ is a synonym for the property $\phi(x)$. With this convention, we can talk about the union or intersection of two classes (since for example $\mathcal{C}(\phi) \cup \mathcal{C}(\psi) = \mathcal{C}((\phi \vee \psi))$.) The statement that the class $\mathcal{C}(\phi)$ is contained in the class $\mathcal{C}(\psi)$ is equivalent to the assertion that $(\phi \rightarrow \psi)$ is logically valid. However, there is no way to give meaning to the assertion that one class is a member of another, so we are not letting Russell's Paradox in by the back door.

Since the first-order language is countable, there are only countably many classes which can be defined in this way. These, however, include such mathematically important examples as the class of all groups. (The formula $\gamma(x)$ defining this class is quite long: it states that x is an ordered pair (g, f), where f is a function from $g \times g$ to g which satisfies the group axioms.)

The union, intersection, and cartesian product of two classes can be defined. For, if the classes $\mathcal{C}_1$ and $\mathcal{C}_2$ are defined by formulae ϕ_1 and ϕ_2 respectively, then

- $\mathcal{C}_1 \cup \mathcal{C}_2$ is defined by $\phi_1 \vee \phi_2$;

- $\mathcal{C}_1 \cap \mathcal{C}_2$ is defined by $\phi_1 \wedge \phi_2$;

- $\mathcal{C}_1 \times \mathcal{C}_2$ is defined by

$$(\exists y)(\exists z)(\phi_1(y) \wedge \phi_2(z) \wedge \pi(x, y, z)),$$

where $\pi(x, y, z)$ is the formula asserting that x is the ordered pair (y, z).

We can extend our notion of classes by allowing the formulae to contain names of sets. Then, in particular, every set is a class: the set a is defined by the formula $x \in a$. A class which is not a set is called a *proper class*.

There is, however, no obvious way to describe the 'power class', as this would involve quantification over first-order formulae.

EXERCISES

6.1   Find a natural bijection between $(A^B)^C$ and $A^{B \times C}$.

6.2   Show that, in ZF, the Axiom of Choice is equivalent to the following assertion:

> If R is an equivalence relation on X, then there is a set of representatives of the R-classes (that is, a subset of X containing exactly one element from each R-class).

6.3   Let A be an infinite set, and let $\mathcal{P}_n A$ denote the set of all n-element subsets of A.

(a) Prove that, if B and C are bijective with A, then $B \cup C$ is bijective with A. [*Note:* B and C are not necessarily disjoint.]

(b) Prove that A is bijective with $\mathcal{P}_n A$.

(c) Prove that A is bijective with the set $\mathcal{P}_{\text{fin}} A$ of all finite subsets of A.

6.4   Prove that there is an operation $\circ$ on any infinite set A such that $(A, \circ)$ is a group. [*Hint:* Exercise 6.3.]

6.5   Prove that every vector space has a basis using Zorn's Lemma: that is, show that maximal linearly independent sets exist, and that such a set must be a basis.

6.6   Prove using Zorn's Lemma that

(a) in a vector space, every linearly independent set is contained in a basis;

(b) in a ring with identity, every ideal is contained in a maximal ideal.

6.7   Show that, in ZFC, there is no set S with the property that $x \in S$ if and only if $|x| = 2$. [*Hint:* Let $T = \bigcup S$ and $U = \{x \in T : x \notin x\}$.]

Show that there is no set G such that $x \in G$ if and only if x is a group. [A *group* is an ordered pair (g, f), where g is a set and $f : g^2 \to g$ is a function satisfying the group axioms.]

6.8   Prove in ZFC that, if R is a partial order on a set X, then there is a total order on X containing R.

6.9   Give an example of a group with no maximal subgroup. Why is it possible to prove in ZFC that any ring with identity has a maximal ideal, but not that any group has a maximal subgroup?

6.10   A *Steiner triple system* on a set X is a set B with the properties

(a) each member of B is a 3-element subset of X;

(b) any two distinct members of X are contained in a unique member of B.

Prove in ZFC that there is a Steiner triple system on any infinite set.

Hint: Well-order both X and the set $X^{(2)}$ of all ordered pairs of distinct elements of X, so that at least $X^{(2)}$ is isomorphic to a cardinal number. Now define the triples by transfinite recursion.

6.11   Let S be the set of natural numbers. Define a relation $\in$ on S as follows. Given a number n, write it (in base 2) as a sum of distinct powers of 2: say

$$n = 2^{m_1} + 2^{m_2} + \cdots,$$

where $m_1, m_2, \ldots$ are all distinct. Now set $x \in n$ if and only if $x = m_i$ for some i.

Show that all the axioms of set theory, except for the Axiom of Infinity, are satisfied.

6.12   Let $f : \mathbb{R} \to \mathbb{R}$ satisfy $f(x + y) = f(x) + f(y)$ for all $x, y \in \mathbb{R}$. Prove that, if f is continuous, then $f(x) = cx$ for some real number c. [*Hint:* Let $c = f(1)$. Prove first that the equation holds for all rational x.]

Prove that in ZFC there are discontinuous solutions to the functional equation $f(x+y) = f(x)+f(y)$. [*Hint:* Choose a *Hamel basis*, a basis for $\mathbb{R}$ as a vector space over $\mathbb{Q}$. Now any linear map from this vector space to itself satisfies the functional equation.]

6.13   Show that there is no set x such that $x \in \bigcup x$. Is there a set x such that $x \in \mathcal{P} x$?

6.14   A *graph* is a set X with an irreflexive and symmetric binary relation S. (A point of the set X is commonly called a *vertex*; an *edge* is a 2-element set $\{x,y\} \subseteq X$ such that $(x,y) \in R$; if $\{x,y\}$ is an edge, we say that x is *joined* to y.)

(a) Prove that there is, up to isomorphism, a unique countable graph $R = (X, S)$ satisfying the following property P:

> If U and V are finite disjoint sets of vertices of R, then there is a vertex z of R which is joined to every vertex in U and to no vertex in V.

[*Hint*: Show that, if the above property P holds in two graphs A and B, then any isomorphism from a finite substructure of A to a finite substructure of B can be extended to any further vertex of A. Now construct an isomorphism from A to B inductively in stages: start with the isomorphism between the empty subsets; at even-numbered stages, extend the isomorphism to a new vertex of A; at odd-numbered stages, extend the inverse of the isomorphism to a new vertex of B. This technique is known as *back-and-forth*. At the end, we have an isomorphism from A to B. So any two countable graphs with property P are isomorphic. We still have to show that a countable graph with property P exists. This can be done directly or deduced from the next part of the question.]

(b) Take any countable model of the Pair Set, Union, Empty Set and Foundation Axioms. Let X be the set of elements of the model, and form a graph on X by 'symmetrising' the membership relation: that is, $(x,y) \in S$ if and only if either $x \in y$ or $y \in x$. Prove that this graph has property P.

[*Note*: This shows that all the countable models of ZF can be obtained by putting directions on the edges of one single countable graph. Moreover, Exercise 6.11 above gives an explicit construction for this graph.]

6.15   Let α be a cardinal number. A theory Σ is said to be α-categorical if any two models of the theory of cardinality α are isomorphic. Show using Exercise 5.2 that the theory of the successor function is α-categorical for any uncountable cardinal number α but not $\aleph_0$-categorical.

6.16   (a) Write down explicitly a formula which defines the class of groups.

(b) Write down a formula which defines the class of ordinals.

(c) State and prove the Principle of Transfinite Induction for the class of ordinals.

6.17   Let $C(n)$ be the assertion that the cartesian product of any family of n-element sets is non-empty. (So $C(2)$ is the statement that Bertrand Russell can always choose his socks.) Prove that $C(2)$ implies $C(4)$.

[*Hint*: Show that if four people play tennis doubles in the three possible combinations, then either one person is always on the winning side, or some person is always on the losing side.]

7
Categories

The standard 'foundation' for mathematics starts with sets and their elements. It is possible to start differently, by axiomatizing not elements of sets but functions between sets. This can be done by using the language of categories and universal constructions.

Saunders MacLane, *Mathematics: Form and Function* [36]

In Chapter 1, we developed various mathematical concepts based on sets. Thus, an ordered pair is a special kind of set, and a function is a set of ordered pairs with certain properties. Natural numbers are particular sets, and from them the standard number systems of mathematics can be developed in familiar ways. More abstractly, a group is an ordered pair whose first element is a set G and whose second element is a function from $G \times G$ to G, satisfying certain axioms. Now such a function is a set of ordered pairs whose first components are in $G \times G$ and whose second components are in G. This is almost but not quite the same thing as a set of ordered triples of elements of G.

If sets are the basic objects of mathematics, then functions are its basic processes. Category theory, which began in a very technical part of mathematics (homology theory), provides the basis of a development in which functions are regarded as fundamental.

Note: In this chapter, we will very often have to compose functions. To make life easier, we will write the composition of f and g (in that order) as fg, rather than $f \circ g$ as in Chapter 1. It is also convenient to write functions on the right of their arguments: thus, the image of x under f will be xf rather than $f(x)$.

Then, indeed, we have $x(fg) = (xf)g$. You are warned that this convention is not universal among category theorists.

The underlying philosophy is that what is important about any class of mathematical structures is the structure-preserving maps between different objects in the class. For example, suppose that our structures are just sets. If $f : X \to Y$ and $g : Y \to Z$ are maps between sets, then there is a composite $fg : X \to Z$. Moreover, f is one-to-one if and only if there is a map $g : Y \to X$ with $fg = 1_X$ (where 1_X is the identity on X); and f is onto if and only if there exists $h : Y \to X$ with $hf = 1_Y$.

Similarly, other set-theoretic notions can be recognized. Here are some more examples. The associative law for groups is usually stated as a law (in the sense of universal algebra), asserting the equality of two expressions $(ab)c$ and $a(bc)$. Another version involves the maps λ_a and ρ_a defined by left and right multiplication by the element a: the associative law asserts that λ_a and ρ_c commute for any $a, c \in G$:

$$b\lambda_a\rho_c = (ab)c = a(bc) = b\rho_c\lambda_a.$$

This version uses a mixture of elements and maps. But the law can be stated using only maps. Let $\mu : G \times G \to G$ be the group operation. If $\alpha_i : G_i \to H_i$ are maps for $i = 1, 2$, then $\alpha_1 \times \alpha_2 : G_1 \times G_2 \to H_1 \times H_2$ is defined coordinatewise. Now the associative law asserts that

$$(1 \times \mu)\mu = (\mu \times 1)\mu$$

as functions from $G \times G \times G$ to G, where 1 is the identity map on G. (The left-hand side maps $(a, b, c) \mapsto (a, bc) \mapsto a(bc)$, and the right-hand side $(a, b, c) \mapsto (ab, c) \mapsto (ab)c$.)

In fact, we can define the cartesian product of two sets using maps. Let X and Y be sets. The cartesian product $X \times Y$ is a set P which has 'projections' f and g to X and Y respectively (by taking the first and second coordinate of each ordered pair). Moreover, if Z is any set and $f' : Z \to X$ and $g' : Z \to Y$ any maps, then there is a map $h : Z \to P$ such that $hf = f'$ and $hg = g'$. (Set $zh = (zf', zg')$.) This property characterizes the cartesian product, up to isomorphism. Moreover, exactly the same properties and characterization hold if we replace sets, maps and cartesian products by groups (or various other kinds of structures), homomorphisms and direct products.

For a final example, a basis X of a vector space V over F is a linearly independent spanning set. However, bases are also characterized (and could be defined) by the following mapping property: any map from X into an F-vector space W can be uniquely extended to a linear map from V to W.

7.1 Categories

These examples give some insight into the viewpoint of category theory. The general definition is as follows.

A *category* consists of the following data:

- a set O of *objects*;

- a set M of *morphisms* or *arrows*;

- a pair of functions, dom (*domain*) and cod (*codomain*), from M to O;

- for each $x \in O$, an *identity morphism* 1_x;

- a partial operation of *composition* on M, the composition of f and g (if it exists) being written fg.

It satisfies the following axioms:

- The composition fg exists if and only if $\operatorname{cod}(f) = \operatorname{dom}(g)$. If this holds, then $\operatorname{dom}(fg) = \operatorname{dom}(f)$ and $\operatorname{cod}(fg) = \operatorname{cod}(g)$.

- If fg and gh are both defined, then $(fg)h = f(gh)$.

- $\operatorname{dom}(1_x) = \operatorname{cod}(1_x) = x$.

- If $\operatorname{dom}(f) = x$ and $\operatorname{cod}(f) = y$, then $1_x f = f = f 1_y$.

We abbreviate the information $\operatorname{dom}(f) = x$ and $\operatorname{cod}(f) = y$ by writing $f : x \to y$.

Part of the philosophy of category theory is that morphisms are more important than objects. In fact, a category can be defined using only the morphisms, the partial composition, and the identity morphisms. We identify the objects with their corresponding identity morphisms. (See Exercise 7.1.)

There are two, quite different, sources of examples of categories. Be careful to distinguish these, though the strength of category theory is that really no distinction needs to be made.

Classes of structures

Let O be a set of mathematical structures of some type. These may be universal algebras of a fixed type (e.g., groups, vector spaces over a given field, lattices). They may also, more generally, be topological spaces, differentiable manifolds, algebraic curves, ...

Let M be the class of all structure-preserving maps between members of O. (For algebras, we could take M to consist of all homomorphisms; for topological spaces, all continuous maps; and so on.) For $f \in M$, we take $\operatorname{dom}(f)$ and $\operatorname{cod}(f)$ to be the usual domain and codomain of f, and take composition to be the usual composition of functions and 1_x to be the identity map on x.

It may also be possible to obtain a category by taking just some of the structure-preserving maps. For example, we could take just the one-to-one maps, or the onto maps; for differentiable manifolds we could take the continuous functions, the differentiable functions, the smooth functions ...

A category of this sort, where the objects are sets (maybe with additional structure) and the morphisms are functions, is called a *concrete category*.

Individual structures

It may surprise you to learn that a group G is an example of a category. Take a single object called $*$ (say), and take G to be the set of morphisms, with $\text{dom}(g) = \text{cod}(g) = *$ for all $g \in G$, and $1_* = 1$, the identity of G. Since there is only one object, any pair of morphisms can be composed.

With this example in mind, we could say that categories form just another type of algebraic object, more general than groups.

But now we have the option of turning the generality on itself. There is a category of all categories!

Categories are more general than groups in two respects: there can be more than one object, and morphisms need not be invertible. Two intermediate classes of structures are obtained by relaxing one or other of these conditions.

A *groupoid* is a category in which any morphism $f : x \to y$ has an inverse $g : y \to x$ (such that $fg = 1_x$ and $gf = 1_y$). An example is obtained by taking any class of structures as objects, and the isomorphisms as morphisms.

A *monoid* is a category with a single object. In other words, it is a set with a (total) operation of composition, satisfying the associative and identity axioms for a group, but not necessarily the inverse axiom. Thus, the endomorphisms of a single structure x (the homomorphisms from x to x) form a monoid.

For convenience, we will in future say 'Let $C = (O, M)$ be a category', meaning that O and M are the sets of objects and morphisms of C. Of course, the notation ignores part of the category structure (the composition and the identities).

Let $C = (O, M)$ be a category. A *subcategory* C' consists of a subset O' of O and a subset M' of M such that $C' = (O', M')$ is a category. An important special case occurs when M' consists of all the morphisms of C whose domain and codomain lie in O'. In this case, C' is called a *full subcategory*. For example, abelian groups form a full subcategory of the category of groups.

As suggested by our remarks in the introduction, we say that the morphism $f : x \to y$ is

- *monic*, or a *monomorphism* if there is a morphism $g : y \to x$ such that $fg = 1_x$;

- *epic*, or an *epimorphism* if there is a morphism $h : y \to x$ such that $hf = 1_y$;

• an *isomorphism* if it is both monic and epic.

In the category of sets and functions, a function is monic if and only if it is one-to-one, and epic if and only if it is onto. (The proof of this uses the Axiom of Choice.) These assertions do not hold in all concrete categories (see Exercise 7.4).

It follows easily from the category axioms that, if f is an isomorphism, then its left and right inverses (g and h in the above definition) are equal. For

$$g = 1_y g = (hf)g = h(fg) = h1_x = h.$$

This morphism is then called the *inverse* of x.

If y is an object for which a monomorphism $f : y \to x$ exists, then y is called a *subobject* of x. Note that this is more general than the algebraic notion of 'subgroup' or 'subring'. For example, in the category of groups and group homomorphisms, a subobject of a group x is a group y which is isomorphic to a subgroup (in the usual sense) of x; that is, a group embeddable in x.

Also motivated by our remarks in the introduction, we define products as follows. Let x and y be objects in a category. Then a *product* of x and y is an object p with a pair of morphisms $f : p \to x$ and $g : p \to y$ with the following property:

for any choice of object z and morphisms $f' : z \to x$ and $g' : z \to y$, there is a unique morphism $h : z \to p$ such that $hf = f'$ and $hg = g'$.

Then the cartesian product is a product in the category of sets, and the direct product is a product in the category of groups.

Theorem 7.1

Any two products of x and y in a category are isomorphic.

Proof

Let (p, f, g) and (p', f', g') be products. Now there are morphisms $f' : p' \to x$ and $g' : p' \to y$; since p is a product, there is a unique morphism $h : p' \to p$ such that $hf = f'$ and $hg = g'$. Similarly, there is a morphism $k : p \to p'$ such that $kf' = f$ and $kg' = g$. Now consider the morphisms kh and 1_p from p to itself. We have $khf = f = 1_p f$ and $khg = g = 1_p g$. By the uniqueness part of the definition of a product, we have $kh = 1_p$. Similarly, $hk = 1_{p'}$. So h and k are inverses, and each is an isomorphism. □

We say that the category C *has products* if any two objects in C have a product.

By 'turning the arrows around', we can define the dual notion. A *coproduct* of x and y is an object q with a pair of morphisms $f : x \to q$ and $g : y \to q$ with the following property:

for any choice of object z and morphisms $f' : x \to z$ and $g' : y \to z$, there is a unique morphism $h : q \to z$ such that $fh = f'$ and $gh = g'$.

The disjoint union is a coproduct in the category of sets. Coproducts are often more difficult to describe: see Exercise 7.6. We say that the category C *has coproducts* if any two objects in C have a coproduct.

7.2 Foundations

As the theory is envisaged, the typical category is something like the 'category of groups', whose objects are all groups and whose morphisms are all group homomorphisms. Unfortunately, groups form a proper class; even trivial groups form a proper class, since there is a group structure on any singleton set. Indeed, any group has a proper class of subobjects. So a more cautious approach is required.

To see the issues, we consider a few categories which will clearly be useful and important:

- **Set**, the category of sets and functions. Russell's Paradox shows immediately that the objects of **Set** form a proper class. This can be handled by the approach to classes that we developed in Section 6.7.

- **Cat**, the category of categories and functors. (We will define functors in the next section; informally, a functor is a 'homomorphism of categories'.) This cannot be handled by the machinery of classes, since not all categories are sets (as we have seen, **Set** is not a set).

- *Functor categories*: the objects of a functor category will consist of all functors from C_1 to C_2, where C_1 and C_2 are categories. If C_1 and C_2 are not sets, this cannot be defined in set theory: a single functor is already a class.

What is even more irritating is the fact that, to an algebraist, there is only one trivial group (up to isomorphism), whereas we have the bewildering situation that there are too many trivial groups to form a set.

As this discussion suggests, problems arise in the first instance because a category may be too 'large' to be a set. Accordingly, we make some definitions.

- A *small category* is a category whose morphisms form a set. (Since there is an injection $x \mapsto 1_x$ from objects to morphisms, the objects of a small category also form a set.) Such categories present no foundational problems. We saw in the last section that any group is a category; in fact, it is a small category.

- A *locally small category* is a category with the property that, for any two objects x and y, the morphisms $f : x \to y$ (that is, with $\text{dom}(f) = x$ and $\text{cod}(f) = y$) form a set. Thus, Set (the category of sets) and Group (the category of groups) are locally small categories, as is the category of small categories.

- A *well-powered category* is one in which, for any object x, there is a set of subobjects of x (objects y for which there is a monomorphism from y to x) such that any subobject of x is isomorphic to one of them. For example, the category of groups is well-powered: for every group is a set (with extra structure), and as representative subobjects of a group x we can take the groups which are subgroups of x in the usual sense (subsets of the set x with the restriction of the operations); these are all contained in $\mathcal{P}\,x$.

We outline three possible approaches to the foundations of category theory.

The first approach might be called 'naïve category theory', by analogy with the naïve set theory of Chapter 1. We proceed directly, taking care to avoid constructions which would lead too close to Russell's Paradox. So small categories are harmless, and locally small or well-powered categories can be handled with the notion of classes. This is the approach most commonly adopted by working mathematicians. For example, one of the main uses of category theory is in ring theory: we get information about a ring R by studying the (locally small, well-powered) category of R-modules.

The second approach was pioneered by Grothendieck. We add new axioms to set theory in order to find a place to work. Grothendieck defined a *universe* to be a set V which is an $\in$-model of set theory: that is, the restriction of the membership relation to V satisfies the Zermelo–Fraenkel axioms. As we have seen, such a set V must have the properties

- V is transitive, that is, if $x \in V$ and $y \in x$ then $y \in V$;

- V contains the set ω of natural numbers;

- if $x, y \in V$ then $\{x, y\} \in V$;

- the power set and union of sets in V are in V;

- the image of a member of V under a function defined by a first-order formula is in V.

As we saw in Section 6.4, if α is an inaccessible cardinal, then V_α is a universe.

Since a universe is a model of set theory, we can do mathematics within V. MacLane proposed adjoining to ZFC the axiom:

There exists a universe.

(As we have seen, this axiom is unprovable in ZFC, since it implies the consistency of ZFC.) Now if V is a universe, we call a category V-*small* if its morphisms form a set in V. Similarly we define V-locally small and V-well-powered categories.

MacLane's Axiom suffices for many purposes, but still limits us, since (for example) the V-small categories form a set, but it is not a member of V. Accordingly, Grothendieck proposed a stronger axiom:

Every set is a member of a universe.

In particular, every universe V is contained in a larger universe V', and the category of V-small categories is V'-small. Further constructions may require a still larger universe V''; but at most a few successive universes suffice for any construction so far proposed.

However, it seems unsatisfactory to have such a plethora of universes. For example, we are forced to replace the single category of trivial groups (which already we felt to be a weakness) with a proper class of categories of trivial groups in the proliferating universes.

The final approach to the foundations is to throw away set theory altogether and build mathematics on the foundations of category theory. Then set theory would appear as Set, the category of sets, just one category among many others. The pioneer of this approach was Lawvere. This attractive program has not yet been carried out satisfactorily, and we will not discuss it further.

For the remainder of this chapter, we will do 'naïve category theory'.

7.3 Functors

The philosophy of category theory is that morphisms carry the essential information about objects. Naturally enough, we next define 'morphisms between categories'.

Let $C = (O, M)$ and $C' = (O', M')$ be categories. A *functor* from C to C' consists of a pair of maps (denoted by the same symbol F) from O to O' and from M to M', satisfying the following conditions:

- $(fg)F = (fF)(gF)$ whenever fg is defined;
- $1_x F = 1_{(xF)}$ for all $x \in O$.

Note that the map on morphisms ('functions') is an important part of a functor, not just an appendage of the map on objects; the name 'functor' is intended to suggest this.

We seem to have left out two very important conditions, namely that F preserves domains and codomains, in the sense that $(\mathrm{dom}(f))F = \mathrm{dom}(fF)$ and $(\mathrm{cod}(f))F = \mathrm{cod}(fF)$ for all $f \in M$. But these follow from the axioms given. For we have

$$1_{\mathrm{dom}(f)}f = f 1_{\mathrm{cod}(f)} = f,$$

so by the first condition,

$$1_{\mathrm{dom}(f)}F \cdot fF = fF \cdot 1_{\mathrm{cod}(f)}F = fF,$$

and the second condition shows that

$$1_{\mathrm{dom}(f)}F = 1_{\mathrm{dom}(f)F}, \qquad 1_{\mathrm{cod}(f)}F = 1_{\mathrm{cod}(f)F}.$$

It then follows that

$$\mathrm{dom}(fF) = \mathrm{dom}(f)F, \qquad \mathrm{cod}(fF) = \mathrm{cod}(f)F.$$

Now we can define the category Cat of categories, whose objects are the categories and whose morphisms are the functors.

Functors occur throughout mathematics. For example, let C be any concrete category. Then the *forgetful functor* from C to Set maps each object to the underlying set and each morphism to the set function. This function simply 'forgets' the structure of the objects of C.

Other functors describe 'partial amnesia'. For example, let C and C' be the categories of rings and abelian groups. Then the partially forgetful functor F maps a ring to its additive group, and a ring homomorphism to the same map (regarded merely as a group homomorphism).

Here are some further examples.

Power set: The power set operation defines a functor from Set to itself, mapping each object x to $\mathcal{P}x$. If $f : x \to y$, we define $\mathcal{P}f : \mathcal{P}x \to \mathcal{P}y$ to be the induced set mapping: that is, $z(\mathcal{P}f) = f[z]$ for $z \subseteq x$. [Note that $\mathcal{P}f$ is not the power set of the set f, but is very closely related to it: every subset of f is a function from a unique subset z of x to y, and the we take the image of this function to be $z(\mathcal{P}f)$.]

Derived group: A different kind of functor maps groups to abelian groups. Let G be a group. The *derived group* G' is the subgroup generated by all commutators $g^{-1}h^{-1}gh$; it is the smallest normal subgroup with abelian factor group. Now there is a functor from groups to abelian groups which maps G to G/G'. Of course, we have to define the action of the functor on morphisms (see Exercise 7.13).

Unit group: The functor Unit, from the category of rings with identity to the category of groups, maps a ring to its group of units. (Check that an identity-preserving ring homomorphism maps units to units and induces a group homomorphism on the group of units.)

General linear group: More generally, for any n, the functor GL_n maps a ring R with identity to the group $GL(n, R)$ of invertible $n \times n$ matrices over R.

Group actions: Recall that any group G is a category with a single object $*$, in which the morphisms are the group elements. What is a functor F from G to the category of sets? $*F$ is a set Ω. For all $g \in G$, gF is a map from Ω to Ω, such that $(g_1 g_2)F = (g_1 F)(g_2 F)$ and $1F$ is the identity map on Ω. This is precisely the definition of a permutation action of G on Ω. So functors from G to sets are permutation representations (actions) of G.

For a categorical form of this example, see Exercise 7.11.

More generally, functions from G to any category C of algebras are actions of G by automorphisms of an algebra in C. For example, if C consists of finite-dimensional vector spaces over F, then a functor from G to C is a representation of G by matrices over F.

7.4 Natural transformations

We move now to the next level in this process. A natural transformation is a homomorphism between functors: with each object, it associates a morphism between the images of the object under the two functors. More precisely, let $F, G : C \to C'$ be functors, where $C = (O, M)$ and $C' = (O', M')$. A *natural transformation* $T : F \to G$ is a function from O to M' with the properties

- for any $x \in O$, $\text{dom}(xT) = xF$ and $\text{cod}(xT) = xG$;

- for any $f \in M$, with $\text{dom}(f) = x$ and $\text{cod}(f) = y$, we have

$$(fF)(yT) = (xT)(fG).$$

Note that $fF : xF \to yF$, $yT : yF \to yG$; and $xT : xF \to xG$, $fG : xG \to yG$, so the composite morphisms on both sides are defined. The condition can be represented by a *commutative square* as follows:

$$
\begin{array}{ccc}
xF & \xrightarrow{\;fF\;} & yF \\
{\scriptstyle xT}\downarrow & & \downarrow{\scriptstyle yT} \\
xG & \xrightarrow{\;fG\;} & yG
\end{array}
$$

(Compare Exercise 7.3.) This means that, if we start from an element of xF and map it to an element of yG by following the arrows along either possible route, the result will be the same (independent of the route taken).

It is probably still not clear what this definition means. In fact, mathematics abounds in important examples. Here are a few.

Determinant: Let C be the category of commutative rings with identity, C' the category of groups. As a small specialization of an earlier example, both Unit (the group of units) and GL_n (the group of invertible $n \times n$ matrices) are functors from C to C'. We claim that det (determinant) is a natural transformation from GL_n to Unit. The first assertion of the definition is that, for any commutative ring R, det is a homomorphism from $GL(n, R)$ to $Unit(R)$: this is a fundamental property of the determinant, namely

$$\det(AB) = \det(A)\det(B).$$

The second property connects this with ring homomorphisms. If $f : R \to S$ is a homomorphism, we have

$$(\det(A))f = \det(Af),$$

where f denotes also the induced maps $GL_n(R) \to GL_n(S)$ and $Unit(R) \to Unit(S)$ (which we might, more consistently, call $GL_n(f)$ and $Unit(f)$).

Group actions: Let G be a group. We saw that a functor from G to the category of sets is just a permutation action of G on a set Ω. A natural transformation between two such functors (actions on sets Ω_1 and Ω_2) is a G-homomorphism between two such actions: that is, a map $T : \Omega_1 \to \Omega_2$ such that $(\alpha g)T = (\alpha T)g$ for all $\alpha \in \Omega_1$.

Double duals: The *dual space* of an F-vector space V is the vector space V' of all linear maps from V to F. Duality is not a functor as we have defined it, since it 'reverses arrows': that is, if $f : V \to W$ is linear, then $f' : W' \to V'$ is defined by

$$v(\phi f') = (vf)\phi$$

for $\phi \in W'$, $v \in V$. (Duality is what is known as a *contravariant functor*.) If D denotes duality, then D^2 is a functor from the category of vector spaces to itself. Also, there is a natural transformation T from the identity to D^2: VT is the map $V \to V''$ under which the image of $v \in V$ is the map $\phi \to v\phi$ from V' to F.

This makes precise the notion that there is a natural embedding of a space into its second dual, independent of any choice of basis. If we want to embed a space in its dual, we must make some choices: the embedding is not 'natural'.

If T is a natural transformation from F to G, and U a natural transformation from G to H, then the composite TU can be defined and is a natural transformation from F to H, by the rule that $x(TU) = (xT)(xU)$ for all objects x. (The right-hand side is the composition of the morphisms $xT : xF \to xG$ and $xU : xG \to xH$.) This operation is just the composition of morphisms in the functor category (see Exercise 7.10).

For example, if T is the natural transformation from the unit group Unit to GL_n defined by mapping an element $u \in \mathrm{Unit}(R)$ to $\mathrm{diag}(u, 1, \ldots, 1)$ (the diagonal matrix with u in the first place and 1 elsewhere), and D is the determinant, then TD is the identity transformation.

EXERCISES

7.1   Given a set M of *morphisms* with a partial composition and a subset I of identities, suppose that the following conditions are satisfied:

- For any $f, g, h \in M$, if fg and gh are defined, then $(fg)h$ and $f(gh)$ are defined and are equal.

- For any $f \in M$, there are unique identities i and j such that if and fj are defined; and $if = fj = f$.

- For $f, g \in M$, fg is defined if and only if there is an identity j such that fj and jg are defined.

- For any identity i, ii is defined.

Prove that M is the set of morphisms of a category.

7.2   Let X and Y be groups. Let P be a group and let $f : P \to X$ and $g : P \to Y$ be group homomorphisms. Suppose that, if Z is any group and $f' : Z \to X$, $g' : Z \to Y$ are homomorphisms, then there is a unique homomorphism $h : Z \to P$ such that $hf = f'$ and $hg = g'$.

Prove that P is isomorphic to the direct product $X \times Y$.

7.3   Let $f : x \to y$ and $f' : x' \to y'$ be morphisms. Define a *commutative square* to be a pair of morphisms $g : x \to x'$ and $h : y \to y'$ such that $fh = gf'$.

Let $C = (O, M)$ be a category, and let Q be the class of commutative squares in C. Show that (M, Q) is a category.

7.4   Show that, in a concrete category, any monomorphism is one-to-one and any epimorphism is onto. Give examples where these implications do not reverse. [*Hint*: Consider cyclic groups of orders 2 and 4.]

7.5   Show that, if (p, f, g) is a product of x and y, then (p, g, f) is a product of y and x.

7.6   (a) Show that, in the category of sets, the 'disjoint union' $(x \times \{0\}) \cup (y \times \{1\})$, with the maps $a \mapsto (a, 0)$ for $a \in x$ and $b \mapsto (b, 1)$ for $b \in y$, is a coproduct of x and y.

   (b) Show that the free product is a coproduct in the category of groups. [You may have to look up the definition of free product: this can be found in [25].]

   (c) Prove that any two coproducts of x and y are isomorphic.

7.7   Let C be a locally small category, x an object of C. Define the (*covariant*) *hom-functor* $\hom(x, \_) : C \to \mathbf{Set}$ as follows:

- the object y is mapped to the set $\hom(x, y)$ of morphisms with domain x and codomain y;

- the morphism $f : y \to z$ is mapped to the function $\hom(x, y) \to \hom(x, z)$ given by composition with f (taking $g \in \hom(x, y)$ to $gf \in \hom(x, z)$).

(The notation suggests that the objects are obtained by substituting objects of C for the blank space). Prove that this really is a functor.

   Define the analogous *contravariant hom-functor* $\hom(\_, x)$.

7.8   Let f_1 and f_2 be morphisms from x to y. An *equalizer* of f_1 and f_2 is a pair (u, e), where u is an object and $e : u \to x$ a morphism, satisfying

- $ef_1 = ef_2$;

- for any morphism $g : z \to x$ such that $gf_1 = gf_2$, there is a unique morphism $h : z \to u$ such that $he = g$.

   (a) Prove that, if (u_1, e_1) and (u_2, e_2) are two equalizers of f_1 and f_2, then u_1 and u_2 are isomorphic.

   (b) Prove that equalizers exist in the category of sets.

(c) Define the dual notion of *coequalizers*, and prove analogues of (a) and (b).

7.9   Show that Grothendieck's Axiom implies that there exists a proper class of universes.

7.10   Let C_1 and C_2 be categories. Show that there is a category C whose objects are the functors from C_1 to C_2 and whose morphisms are the natural transformations between them. (This is the *functor category* $C_2^{C_1}$.)

7.11   It follows from the preceding exercise and the example in Section 7.3 that, for any group G, the objects of the functor category Set^G are the G-spaces (or sets with an action of G). What are the morphisms of this category?

7.12   (a) Construct a functor $F : \mathsf{Set} \to \mathsf{Group}$ such that, for any set x, the group xF is the free group on the generating set x.

(b) If G denotes the forgetful functor from Group to Set, find a natural transformation from the identity functor on Set to the functor FG.

7.13   Let $\theta : G \to H$ be a homomorphism of groups. Prove that $G'\theta \leq H'$. Hence show that θ induces a unique homomorphism $\theta^* : G/G' \to H/H'$.

Hence show how to define a functor from groups to abelian groups which maps G to G/G' for any group G.

7.14   Why is there no natural way to define a functor from groups to abelian groups which maps G to $Z(G)$ (the centre of G)?

7.15   A *preorder* is a reflexive and transitive relation on a set.

(a) Show that any preordered set (X, P) is a category, with object set X and morphism set P, with $\mathrm{dom}(x, y) = x$ and $\mathrm{cod}(x, y) = y$ for all $(x, y) \in P$, and $1_x = (x, x)$.

(b) Show that a category is a preorder if and only if there is at most one morphism with any given domain and codomain.

8

Where to from here?

This chapter rounds off the book with a few remarks on the philosophy of mathematics and some suggestions for further reading.

8.1 Philosophy of mathematics

'If two and two can be four then they actually are four, you can only perceive it, you have no part in making it happen by writing it down in numbers or telling it out in pebbles.'

Russell Hoban, *Pilgermann* [20]

... mathematical knowledge ... is, in fact, merely verbal knowledge. '3' means '2 + 1', and '4' means '3 + 1'. Hence it follows (though the proof is long) that '4' means the same as '2+2'. Thus mathematical knowledge ceases to be mysterious.

Bertrand Russell, *History of Western Philosophy* [41]

The most important question that the philosophy of mathematics has to address is the eerie stability of the subject. Physics had a major conceptual revolution in the seventeenth century, and two more in the early twentieth century; the story in biology is similar. In these sciences, fundamental premises of the old theories were contradicted by the new (though the predictions of the theories for events on a human scale were similar). In mathematics, by contrast, we still teach Pythagoras' Theorem as it was proved two and a half millennia ago.

To understand how the human activity of mathematics has such permanence, we need to distinguish between the creation of mathematics and the created product. Mathematicians, like workers in any creative endeavour, are guided by intuition, guesswork, hard slog, and lucky chances, while proving their theorems. But mathematicians 'cover their tracks'; the proof as it appears in the journal shows little evidence of its creation. (We teach students to analyze a problem by working back from the conclusion, and then write out the steps in the reverse order; this is a simple example of such a trick.)

Also, to say that mathematical knowledge is stable is not to say that mathematics is immune to the errors and trends of fashion which occur in other subjects. In the late nineteenth century, descriptive invariant theory was extremely popular. It was killed off by the axiomatic approach, and virtually ignored for much of the twentieth century, until various new areas of mathematics such as coding theory created a demand for the old results. But, though these results had been forgotten, they had not been rejected or overturned.

We also need to distinguish carefully between the formal definition of a mathematical object (what it is) and its properties (what it does). Historically, the properties come first. Group theory was nearly a hundred years old before the modern definition of a group was given. Previously, a group was a set of transformations, closed under composition and inversion and containing the identity transformation. Cayley's Theorem shows that any group (in the axiomatic sense) can be represented as such a set of transformations. So the properties remain the same when the definition changes. A more extreme example involves the natural numbers. Nobody thinks of the number 4 as the set

$$\{\varnothing, \{\varnothing\}, \{\varnothing, \{\varnothing\}\}, \{\varnothing, \{\varnothing\}, \{\varnothing, \{\varnothing\}\}\}\},$$

but everybody agrees that $2 + 2 = 4$. It seems clear that, even if great changes in the foundations of mathematics lead us to an entirely different definition of the number 4 in future, this property of 4 will still be accepted. Traditional philosophy of mathematics concerns itself with whether the proposition $2+2 = 4$ is a necessary truth, or if not, what its status is.

Recent research (reported in Dehaene [13]) suggests that humans are born with a number instinct, somewhat like the language instinct which has been postulated by Chomsky's school of linguistics. Clearly, babies have an instinct for what numbers do, not for the set-theoretic definition!

A related point is the 'unreasonable effectiveness' of mathematics as a description of the real world. The theories of physics are formulated in mathematical language; the calculations performed by physicists enable them to predict the results of experiments with extraordinary accuracy.

I once heard a lecture by a category theorist who, quoting Heraclitus and Hegel, regarded mathematics as a dialectic process, the thesis and antithesis corresponding to a functor and its adjoint between two categories. (The pre-Socratic philosopher Heraclitus said (Fragment 45),

> Men do not understand how that which differs with itself is in agreement: harmony consists of opposing tension, like that of the bow and the lyre.

The arrows from his bow suggest those between objects in a category. Hegel's dialectic postulates that one concept (thesis) inevitably generates its opposite (antithesis); their interaction leads to a new concept (synthesis), which in turn becomes the thesis of a new triad.) This metaphor describes the creation and development of a branch of mathematics.

Lakatos [32] eloquently describes the process of mathematical discovery and research, the need to test proofs against examples to discover the weak links and hidden assumptions, and the way that premature optimism can be deflated. For anyone who believes in the infallibility of mathematicians, this is key reading matter!

Many mathematicians have speculated in print about the creative process in mathematics. The most detailed investigations are those of Poincaré [40] and Hadamard [17]. Littlewood [34] gives practical rules of work for a mathematician; van der Waerden [44] gives a factual, and Knuth [31] a fictional, account of mathematical discovery.

In the late nineteenth and early twentieth century, developments such as non-Euclidean geometry and the paradoxes in set theory led to the appearance of a number of schools of philosophy of mathematics. Heyting [19] gives an account of this, couched as a Socratic dialogue between representatives of the schools (somewhat biased, since he was one of the founders of the school of Intuitionism, and the book is an introduction to the tenets of this school). Nowadays, mathematicians seem more comfortable with their subject, and less inclined to form themselves into schools. One small but still active school is *constructivism*, which asserts that the only valid proof is an explicit construction; 'proof by contradiction' is not valid. See Bishop and Bridges [5] for an exposition of the views of this school.

A story compares researchers in the foundations of mathematics to a family of spiders who lived in a disused dungeon of an old castle, spinning their webs across the underground space. Once a year, a lackey would come down and clear away all the cobwebs. The spiders would cower in the corners, convinced that without this support the castle would surely fall down.

8.2 Further reading

A number of books give a different account of what has been treated here: these include Enderton [15] for set theory, Hamilton [18] for logic, and MacLane [35] or McLarty [37] for category theory. It is always useful to get somebody else's view on the subject; if for no other reason, you may find that one author explains the views another takes for granted. For more on the background discrete mathematics and algebra, see Biggs [6] or Cameron [9]. All of Chapters 1–5, and more, are treated in the remarkable little book *What is Mathematical Logic?*, by John Crossley *et al.* [10]. It captures the essence with almost no technical detail, and also treats the independence of the Axiom of Choice and recursive function theory, and includes a historical survey.

In set theory, a very insightful book which goes further in several directions is Devlin's *The Joy of Sets* [14]. This book includes an account of two completely different constructions of models of set theory (Gödel's *constructible universe*, in which the Axiom of Choice and the Continuum Hypothesis are true, and *forcing*, invented by Paul Cohen, which can be used to construct models in which these statements are false). It also includes a chapter on set theory without the Axiom of Foundation. This has recently become popular as a way of modelling recursive processes in computer science and elsewhere, and a more detailed account is given by Barwise and Moss [4].

A definitive account of model theory is given by Hodges [21]. As we saw in Chapter 5, this subject has many connections with other parts of mathematics. For models of Peano arithmetic, see Kaye [28]; for the relation with automorphisms, see Kaye and Macpherson [29]. An important area not touched on here is *recursion theory*, which is closely connected both with Gödel's incompleteness theorems and with the theory of computability in computer science: see Cutland [11]. Hofstadter [22] gives an entertaining discussion of Gödel's Theorem and the issues surrounding it.

Category theory is, as we have suggested, both a foundational subject and an important tool in algebra. From the first point of view, McLarty [37] will lead you to the study of toposes, where you could read the important book by Johnstone [27]. In the other direction, an area of mathematics in which category theory is an indispensable language is ring theory: we now routinely get information about a ring by studying various categories of modules for that ring. This approach is described in Adkins and Weintraub [1]. Category theory is also a useful language in computer science: see Pierce [39].

EXERCISES

8.1 'Mathematics is merely verbal', according to Russell. Why then do so many people agree with Hoban's assertion?

8.2 Is mathematics discovered or invented? If it is discovered, what is it before its discovery? If it is invented, why could not contradictory propositions be invented too?

8.3 'A proof only becomes a proof after the social act of "accepting it as a proof".' (Yu. I. Manin). Is this true, and if so, does it subvert traditional ideas of mathematical rigor?

8.4 Compare Euclid's and Hilbert's axioms for geometry. Are these two authors studying the same subject?

8.5 Can a proof involving a computation be valid mathematics? Is your conclusion affected if the computer is executing a randomized algorithm with a probability 10^{-1000} of producing an incorrect answer? Also, would you distinguish between an explicit construction (checkable by hand) and a non-constructive proof (such as an exhaustive search)?

You should consider some particular instances of proofs involving computers, such as the Four-Colour Theorem (Appel and Haken [2]) and the non-existence of a projective plane of order 10 (Lam *et al.* [33]).

8.6 The Axiom of Choice is neither provable nor disprovable in ZF. Are there any circumstances which could lead to its acceptance or rejection by the mathematical community?

8.7 Would a mathematics based on process rather than structure (whether founded on category theory or something else) be significantly different from our present mathematics?

8.8 Invstigate the contributions of Aristotle, William of Ockham, and Leibniz to the development of logic.

8.9 Does modern set theory solve the philosophical problem of universals?

8.10   What is *intuitionistic logic*? Can it be described as a formal system? If so, is there a Soundness and Completeness Theorem, or a decision procedure? How is it related to classical logic?

Solutions to selected exercises

1.1 Let x be any set. Then for any set z, the implication $(z \in \varnothing) \Rightarrow (z \in x)$ is true, since $(z \in \varnothing)$ is false; thus $\varnothing \subseteq x$.

1.2 (a) True. If $x \in X$, then $\{x\} \in \mathcal{P} X$, and so $x \in \bigcup \mathcal{P} X$. Conversely, if $x \in \bigcup \mathcal{P} X$, then $x \in Y$ for some $Y \in \mathcal{P} X$; then $Y \subseteq X$, and so $x \in X$.

(b) False. If $X = \{\{1\}\}$, then $\bigcup X = \{1\}$, and $\mathcal{P} \bigcup X = \{\varnothing, \{1\}\}$.

It is true that $X \subseteq \mathcal{P} \bigcup X$. For take $x \in X$. Then every element of x is contained in $\bigcup X$, so $x \subseteq \bigcup X$, that is, $x \in \mathcal{P} \bigcup X$.

(c) From (a) and (b) we see that these two sets are not equal but $\bigcup \mathcal{P} X \subseteq \mathcal{P} \bigcup X$.

(d) False. If X has m elements and Y has n elements, then $\mathcal{P}(X \times Y)$ has 2^{mn} elements while $\mathcal{P} X \times \mathcal{P} Y$ has 2^{m+n} elements. If $m = n = 1$, then the second is greater than the first, while if $m = n = 3$, the first is greater than the second. So neither of the sets can be a subset of the other for all choices of X and Y.

(e) False. If $X = \{1\}$ and $Y = \{2\}$, then the set $\{1, 2\}$ belongs to $\mathcal{P}(X \cup Y)$ but not to $\mathcal{P} X$ or $\mathcal{P} Y$.

It is true that $\mathcal{P} X \cup \mathcal{P} Y \subseteq \mathcal{P}(X \cup Y)$. For every subset of either X or Y is a subset of $X \cup Y$.

1.4 (a) No: $(1, 2, 1) = (2, 1, 2)$.

(b) Yes: if $((x_1, y_1), (y_1, z_1)) = ((x_2, y_2), (y_2, z_2))$, then $(x_1, y_1) = (x_2, y_2)$ and $(y_1, z_1) = (y_2, z_2)$, so $x_1 = x_2$, $y_1 = y_2$ and $z_1 = z_2$.

(c) No: $(1, 1, 2) = (1, 2, 2)$.

1.6 let f_g be the function $x \mapsto \mu(x, g)$. To show that f_g is injective, suppose that $f_g(x) = f_g(y)$, that is, $\mu(x, g) = \mu(y, g)$. Then

$$x = \mu(x, 1) = \mu(\mu(x, g), g^{-1}) = \mu(\mu(y, g), g^{-1}) = \mu(y, 1) = y.$$

To show that f_g is surjective, take $z \in X$, and let $x = \mu(x, g^{-1})$. Then

$$z = \mu(z, 1) = \mu(\mu(z, g^{-1}), g) = \mu(x, g) = f_g(x),$$

as required.

1.8 (a) Let f be injective. Define $g : Y \to X$ as in Theorem 1.8. Then $f \circ g = i_X$.

Conversely, suppose that $f \circ g = i_X$, and let $f(x_1) = f(x_2)$. Then $x_1 = g(f(x_1)) = g(f(x_2)) = x_2$. So f is injective.

(b) Let f be surjective. For each $y \in Y$ choose $x \in X$ with $f(x) = y$ (here the Axiom of Choice is used), and define $h : Y \to X$ by $h(y) = x$. Then $h \circ f = i_Y$.

Conversely, suppose that $h \circ f = i_Y$, and choose $y \in Y$. Then f maps $h(y)$ to y; so f is surjective.

1.12 (a) Choose $x_0 \in X$. If it is not minimal, choose $x_1 < x_0$, and so on. This descending chain has no repetitions, so must terminate in a minimal element.

(b) The unique minimal element of a finite totally ordered set is its least element. Given two finite totally ordered sets of the same size, match their least elements, and proceed by induction.

(c) Let R be a non-strict partial order on X, where X is finite. Suppose that R is not total, so that there exist incomparable elements $a, b \in X$. Let

$$R^+ = R \cup \{(x, y) \in X \times X : (x, a), (b, y) \in R\}.$$

Case analysis shows that R^+ is a partial order containing R. After finitely many steps of this kind, we reach a total order.

1.14 (a) We show that $\mathbb{N}^n$ is countable by induction on n. The assertion is clearly true for $n = 1$. Suppose that $\mathbb{N}^n$ is countable. Then

$$\mathbb{N}^{n+1} = \mathbb{N}^n \times \mathbb{N}$$

is the cartesian product of two countable sets, so is countable.

(b) The set of n-tuples of elements of X is countable, by part (a). So X^* is the union of countably many countable sets (namely X^n for each natural number n), so is countable.

(c) A polynomial equation of degree n is specified by $n + 1$ coefficients. By part (b), the set of equations is countable. But an equation of degree n has at most n real roots. So the set of algebraic numbers is the union of countably many countable sets, hence countable.

If the set of transcendental numbers were countable, then the set of all real numbers would be the union of two countable sets, whence countable, which it is not. So the set of transcendental numbers is uncountable.

This is Cantor's proof of the existence of transcendental numbers: an uncountable set cannot be empty!

1.16 Follow the hint, as detailed. The function f defined in this way is certainly one-to-one. It is order-preserving: for the construction ensures that the order relation holding between $f(x_n)$ and each $f(x_i)$ for $i < n$ is the same as that between x_n and x_i. The difficult part is to show that f is onto.

Suppose, for a contradiction, that f is not onto. Let q_m be the rational with smallest index which is not in the image of f. Then $q_0, \ldots, q_{m-1}$ are all in the image of f. Choose n such that

$$\{q_0, \ldots, q_{m-1}\} \subseteq \{f(x_0), \ldots, f(x_{n-1})\}.$$

Now q_m lies in one of the $n + 1$ intervals into which the line is divided by the n points $f(x_0), \ldots, f(x_{n-1})$; say $q_m \in (f(x_i), f(x_j))$. (The argument is similar if q_m is in one of the semi-infinite intervals at either end.) The corresponding interval (x_i, x_j) in X is non-empty (by the denseness of X; or, if we are dealing with a semi-infinite interval, by the unboundedness of X). Let x_r be the point of X in this interval with smallest index. When we come to define $f(x_r)$, we find $x_r \in (x_i, x_j)$, so we must choose $f(x_r)$ to be the rational with smallest index in $(f(x_i), f(x_j))$. But this is q_m, since all of $q_0, \ldots, q_{m-1}$ have already been chosen. Thus, $f(x_r) = q_m$, contrary to the assumption that q_m is not in the image of f.

Thus f is onto, and so is an order-isomorphism.

1.18 First we make the following observation:

$$f(\{x - n + 1, x - n + 2, \ldots, x\}) = \binom{x + 1}{n} - 1.$$

For this, we use the standard identity

$$\binom{y}{r - 1} + \binom{y}{r} = \binom{y + 1}{r}$$

for binomial coefficients. Now, if we add one to the left-hand side of the first equation, the first two terms become

$$\binom{x - n + 2}{1} + \binom{x - n + 2}{2} = \binom{x - n + 3}{2};$$

this term then adds to the next term $\binom{x-n+3}{3}$ to give $\binom{x-n+4}{3}$; the process continues like a row of dominoes until we have a single term $\binom{x+1}{n}$.

Now we show that f is a bijection by showing that there is a unique solution $(x_0, \ldots, x_{n-1})$ of the equation $f(\{x_0, \ldots, x_{n-1}\}) = N$ with $x_0 < \ldots < x_{n-1}$. The proof is by induction on n. Suppose that

$$\binom{y}{n} \le N < \binom{y+1}{n}.$$

Then we must choose $x_{n-1} = y$, since if $x_{n-1} = x$ then the maximum possible value of $f(\{x_0, \ldots, x_{n-1}\})$ would be $\binom{x+1}{n} - 1$, by our earlier calculation. Then we have to choose $x_0, \ldots, x_{n-2}$ so that

$$f(\{x_0, \ldots, x_{n-2}\}) = N - \binom{y}{n}.$$

By the inductive hypothesis, there is a unique solution; and this solution satisfies $x_{n-2} < y$, since

$$N - \binom{y}{n} < \binom{y+1}{n} - \binom{y}{n} = \binom{y}{n-1}.$$

The result is proved.

The induction begins since for $n = 1$ the function f is simply given by $f(\{x\}) = x$.

1.19 Assume (a) (the Axiom of Choice), and let P be a partition of X. Let F be the identity function on P. Then $F(p) = p \ne \varnothing$ for all $p \in P$. Let f be a choice function, and $Y = \{f(p) : p \in P\}$. Then, for every $p \in P$, $Y \cap p = \{f(p)\}$.

Conversely, assume (b), and let F be any function on X such that $F(X) \ne \varnothing$ for all $x \in X$. Let $Z = \{(x, y) : y \in F(x), x \in X\}$. Now $P = \{\{x\} \times F(x) : x \in X\}$ is a partition of Z. Choose a set Y meeting every set of this partition in just one point. Thus, for each $x \in X$, there is a unique $y \in F(x)$ such that $(x, y) \in Y$. Now Y is itself a choice function for F.

1.20 Let g be a surjection from Y to X. Let F be the function from X to $\mathcal{P}Y$ given by

$$F(x) = \{y \in Y : g(y) = x\}.$$

By assumption, $F(x) \ne \varnothing$ for all $x \in X$. Let f be a choice function for F. Then $f(x) \in F(x) \subseteq Y$ for all $x \in X$, that is, f is a function from X to Y. Now clearly if $x_1 \ne x_2$, then $F(x_1)$ and $F(x_2)$ are disjoint, so $f(x_1) \ne f(x_2)$; so f is an injection.

2.1 This involves checking the axioms, case-by-case. For the ordinal sum, we simplify the notation by using X and Y in place of $X \times \{0\}$ and $Y \times \{1\}$, assuming that X and Y are disjoint.

(a) Call the three clauses of the definition (1), (2), (3).

Irreflexivity: $z < z$ cannot be as a result of (3); if $z \in X$ then $z \not< z$ since X is ordered; and if $z \in Y$ then $z \not< z$ since Y is ordered.

Trichotomy: Suppose that $z_1 \neq z_2$. If $z_1, z_2 \in X$, then one of $z_1 < z_2$ and $z_2 < z_1$ holds since X is totally ordered. Similarly if $z_1, z_2 \in Y$. If, say, $z_1 \in X$ and $z_2 \in Y$, then $z_1 < z_2$ by (3).

Transitivity: Suppose that $z_1 < z_2$ and $z_2 < z_3$. If $z_1, z_2, z_3 \in X$, then $z_1 < z_3$ since X is ordered. So assume that at least one of the points is in Y. Similarly, we can assume that at least one is in X. Without loss of generality, $z_2 \in X$. Then $z_1 \in X$ and $z_3 \in Y$, so $z_1 < z_3$.

(b) Call the two clauses (1) and (2).

Irreflexivity: Clear.

Trichotomy: Suppose that $z_1 = (x_1, y_1) \neq z_2 = (x_2, y_2)$. If $y_1 \neq y_2$, then without loss $y_1 < y_2$, so $z_1 < z_2$ by (1). If $y_1 = y_2$, then $x_1 \neq x_2$ (property of ordered pairs); without loss, $x_1 < x_2$, and so $z_1 < z_2$ by (2).

Transitivity: Suppose that $z_1 < z_2$ and $z_2 < z_3$, where $z_i = (x_i, y_i)$. If y_1, y_2, y_3 are not all equal then (by considering four sub-cases) $y_1 < y_3$, so $z_1 < z_3$ by (1). Otherwise, the ordering of the z_i is the same as that of the x_i by (2), and transitivity for X implies the result.

Now suppose that X and Y are well-ordered.

(a) Let $S \subseteq X \cup Y$, $S \neq \varnothing$. If $S \cap X \neq \varnothing$ then, since X is well-ordered, there is a least element s of $S \cap X$. By (1), $s < y$ for all $y \in S \cap Y$; so s is the least element of S. On the other hand, if $S \cap X = \varnothing$ then $S \subseteq Y$, and so S has a least element since Y is well-ordered.

(b) Let $S \subseteq X \times Y$, $S \neq \varnothing$. Let

$$U = \{y \in Y : (\exists x \in X) \text{ with } (x, y) \in S\}.$$

Then $U \neq \varnothing$, so U has a least element u. Now let

$$T = \{x \in X : (x, u) \in S\}.$$

Then T has a least element t. We claim that (t, u) is the least element of S. If $(x, y) \in S$, $(x, y) \neq (t, u)$, then either $y \neq u$ (whence $u < y$, and $(t, u) < (x, y)$ by (1)), or $y = u$, $x \neq t$ (whence $t < x$, and $(t, u) < (x, y)$ by (2)).

2.2 If X is well-ordered, then X^2 is well-ordered: take it to be the lexicographic product of the ordered set X with itself. By induction, X^n is well-ordered for all $n \geq 1$. Now X^0 has just one element, namely the empty sequence. Now take the ordered sum of the well-ordered sets X^n for all n; that is, if $s \in X^n$ and $t \in X^m$, put $s < t$ if either $n < m$, or $n = m$ and $s < t$ as element of X^n.

Suppose that $a, b \in X$ with $a < b$. Then, in the dictionary order on X^*, we have the infinite decreasing sequence

$$b > ab > aab > aaab > aaaab > \cdots$$

2.6 All the proofs are by induction. Here is the working for (a).
Suppose that $\gamma = 0$. Then

$$(\alpha + \beta) + 0 = \alpha + \beta = \alpha + (\beta + 0).$$

Suppose that $\gamma = s(\delta)$, and assume that $(\alpha + \beta) + \delta = \alpha + (\beta + \delta)$. Then

$$\begin{aligned}
(\alpha + \beta) + s(\delta) &= s((\alpha + \beta) + \delta) \\
&= s(\alpha + (\beta + \delta)) \\
&= \alpha + s(\beta + \delta) \\
&= \alpha + (\beta + s(\delta)).
\end{aligned}$$

Finally, suppose that γ is a limit ordinal, and that $(\alpha + \beta) + \delta = \alpha + (\beta + \delta)$ for all $\delta < \gamma$. Then

$$\begin{aligned}
(\alpha + \beta) + \gamma &= \bigcup_{\delta < \gamma} (\alpha + \beta) + \delta \\
&= \bigcup_{\delta < \gamma} \alpha + (\beta + \delta) \\
&= \alpha + \bigcup_{\delta < \gamma} (\beta + \delta) \\
&= \alpha + (\beta + \gamma).
\end{aligned}$$

2.8 (a) For each $x \in X$, the function f given by $f(i) = x$ for all $i \in I$ is a choice function. This shows that the cartesian product is at least as large as X.

(b) Let x_i be the least element of X_i. Then the function f given by $f(i) = x_i$ for all $i \in I$ is a choice function.

2.10 (a) The proof is by induction. The conclusion is clear for a limit ordinal, so suppose that α is a successor ordinal, say $\alpha = s(\beta)$. By the inductive hypothesis, $\beta = \lambda + m$, where λ is a limit ordinal and m a natural number. Now

$$\alpha = \beta + 1 = (\lambda + m) + 1 = \lambda + (m + 1),$$

which is of the required form.

(b) Let λ be a limit ordinal. By induction and part (a), every ordinal smaller than λ can be written in the form $\omega \cdot \beta + n$ for some ordinal β and natural number n. Let α be the set of all the ordinals β which occur in such expressions. Then we have $\beta < \alpha$, so $\omega \cdot \beta + n < \omega \cdot \alpha$; thus, $\lambda \leq \omega \cdot \alpha$. On the other hand, every ordinal less than $\omega \cdot \alpha$ has the form $\omega \cdot \beta + n$ for some $\beta < \alpha$; so $\omega \cdot \alpha \leq \lambda$, and we have equality.

2.12 The set of countable ordinals is an ordinal, since every section of it is a countable ordinal. It cannot be a countable ordinal, else it would be smaller than itself. So it is uncountable.

By Exercise 1.17, every countable ordered set (and in particular every countable ordinal) is isomorphic to a subset of the ordered set $\mathbb{Q}$. So $\mathbb{Q}$ has uncountably many non-isomorphic subsets.

3.1 An easy induction shows that the theorems of the MU-system satisfy the first condition, and we proved in Section 3.1 that they also satisfy the second condition.

So let ϕ be a formula which satisfies these two conditions. Let x and y be the numbers of occurrences of I and U in ϕ. Let 2^k be the smallest power of two such that $2^k > x + 3y$ and 2^k is congruent to $x + 3y \bmod 2$. (By assumption, $x + 3y$ is not divisible by 3; and powers of 2 are alternately congruent to 1 and 2 mod 3.)

Now start with MI. Applying Rule 2 k times gives M followed by 2^k Is. If $x + 3y$ is odd, apply Rule 1 to add a U. Now apply Rule 3 repeatedly to replace the last $2^k - (x + 3y)$ Is by $(2^k - (x + 3y))/3$ Us, and Rule 4 to delete these Us (and the extra one if $x + 3y$ is odd). This leaves M followed by $x + 3y$ Is, from which Rule 3 applied y times in the appropriate places yields ϕ.

3.3 You may have found this exercise very difficult. It includes all the theorems that we needed in the proof of the Completeness Theorem; so, once it is proved, it is never again necessary to devise a proof in propositional logic; simply compute a truth table.

The arguments given are skeletons of formal proofs.

(a) From the set $\{(\neg \psi)\}$, we can deduce

$$((\neg \theta) \rightarrow (\neg \psi))$$

(using (A1)), and then

$$(\psi \rightarrow \theta)$$

(using (A3)). The result now follows from the Deduction Theorem. This result tells us that, if we can deduce both ψ and $(\neg \psi)$, then we can deduce any formula (from the same hypotheses).

(b) Suppose that both ψ and $(\neg \psi)$ have been deduced from $\Sigma \cup \{(\neg \phi)\}$. Let α be any instance of an axiom. By (a), we can deduce $(\neg \alpha)$ from the same hypotheses. Hence, by the deduction theorem, from Σ we can deduce

$((\neg\phi) \to (\neg\alpha))$, and hence also $(\alpha \to \phi)$ (by (A3)). Since α is an axiom, we can deduce ϕ.

(c) (i) From $(\neg\phi)$, $((\neg\phi) \to \psi)$ and $((\neg\phi) \to (\neg\psi))$, we can deduce both ψ and $(\neg\psi)$. By (b), we can deduce ϕ from the second and third hypotheses. The result now follows from two applications of the Deduction Theorem.

(ii) From $(\neg\phi)$ and $(\neg(\neg\phi))$ we can immediately deduce a proposition and its negation. By (b), we can deduce ϕ from $(\neg(\neg\phi))$. Now use the Deduction Theorem.

(iii) From $(\neg(\neg\phi))$, $(\phi \to \psi)$ and $(\neg\psi)$, we obtain ψ and its negation (using (ii)). So from the second and third hypotheses we get $(\neg\phi)$. Now use the Deduction Theorem.

(iv) Using (iii), from $(\phi \to \psi)$ and $((\neg\phi) \to \psi)$ we get $((\neg\psi) \to (\neg\phi))$ and $((\neg\psi) \to \phi)$, and hence ψ (using (i)). Now use the Deduction Theorem.

(v) From ψ we get $((\psi \to \theta) \to \theta)$, and hence $((\neg\theta) \to (\neg(\psi \to \theta)))$. Finish as usual.

3.8 Represent the truth values T and F by 0 and 1 respectively, and let y_i be the value 0 or 1 corresponding to $v(p_i)$. (These values are taken in the binary field $\mathbb{Z}/(2)$.) If we let $f(\phi)$ be the value corresponding to $v(\phi)$, then we find that

$$f((\neg\phi)) = f(\phi) + 1, \qquad f((\phi \leftrightarrow \psi)) = f(\phi) + f(\psi).$$

Hence by induction

$$f(\phi) = a_0 + a_1 y_1 + \cdots + a_n y_n,$$

where $a_i = x_i(\phi)$. Now (a) follows, since ϕ is a tautology if and only if $f(\phi) = 0$.

(b) If ϕ is not a tautology but $x_0(\phi)$ is even, then $f(\phi)$ defines a linear map from $(\mathbb{Z}/(2))^n$ to $\mathbb{Z}/(2)$. Its kernel is a subspace of codimension one, and so contains half of the vectors. Adding 1 (that is, negating) has the effect of interchanging the values 0 and 1 taken by a formula, and again half the valuations will map it to zero. Now, given ϕ and ψ, the valuations v for which $v(\phi) = v(\psi)$ are just those for which $v((\phi \leftrightarrow \psi)) = T$, which are half of all the valuations unless ψ is equivalent to ϕ or $(\neg\phi)$.

(c) Suppose that s errors occur. Then the received sequence differs from the correct one in s places. Also, since any two transmitted sequences differ in at least 2^{n-1} places, the received sequence differs from any other sequence in at least $2^{n-1} - s$ places. So, if $s < 2^{n-2}$, we can recognise the transmitted sequence as the unique one nearest to the received sequence.

3.10 The proof of soundness (that the three rules preserve truth) is a simple truth table argument. For example, consider the Contradiction Rule. We have to show that, if a valuation v satisfies $v(((\neg\phi) \to \psi)) = v(((\neg\phi) \to (\neg\psi))) = T$, then $v(\phi) = \mathsf{T}$. If the conclusion were false, then either $v(\psi)$ or $v((\neg\psi))$ would be F; and $v((\neg\phi)) = \mathsf{T}$, so the hypothesis is contradicted.

To prove completeness, it is enough to prove axioms (A1)–(A3). Here is the proof of (A3); the other two are similar but easier. From the hypotheses $\{((\neg\phi) \to (\neg\psi)), \psi, (\neg\phi)\}$ we can infer ψ. By the Deduction Theorem (which is a basic rule here, not a metatheorem!), from the set $\{((\neg\phi) \to (\neg\psi)), \psi\}$, we infer $((\neg\phi) \to \psi)$; and also, of course, $((\neg\phi) \to (\neg\psi))$. Now, by the Contradiction Rule, from the same hypotheses, we can infer ϕ. Two applications of the Deduction Theorem complete the proof.

3.11 There is a formal system in which all tautologies are theorems: we simply take all tautologies as axioms. (Since tautologies are recognisable by the mechanical truth table method, this does indeed satisfy our definition of a formal system.) However, there is no such formal system in which all logical consequences of Σ are provable from Σ, for all sets Σ of formulae: without rules of inference we can prove nothing except axioms and members of Σ.

4.1 The mathematical proof works as follows: suppose that $x^2 = 1$ for all $x \in G$ (this is shorthand for $\mu(x, x) = \epsilon$). Then, for all x and y, we have $xyxy = 1$. Multiplying on the left by x and on the right by y, using the associative law and the facts that $xx = yy = 1$, we obtain $yx = xy$. Your job is to translate this proof into first-order language.

4.3 The closure law is crucial for checking whether a non-empty subset of a group is a subgroup.

5.1 'Only if' is clear; we prove 'if'. So let G be a graph in which every finite subgraph can be coloured with r colours. Take a first-order language with a constant symbol c_i corresponding to each vertex v_i of G, and unary relation symbols $R_1, \ldots, R_r$. Let Σ be the set of sentences asserting that, for any x, exactly one of $R_1(x)$, ..., $R_r(x)$ holds; $c_i \neq c_j$ for all distinct i, j; and for $k = 1, \ldots, r$ and each edge $\{v_i, v_j\}$ of G, the sentence $\neg(R_k(c_i) \wedge R_k(c_j))$.

Given any finite subset Σ_0 of Σ, let G_0 be the graph consisting of all vertices mentioned in the sentences of the last kind in Σ_0. Take a colouring of G_0 with r colours, and use this to define the relations R_i on the vertices of G_0. Let all other vertices of G satisfy R_1. (We interpret c_i as v_i.) This is a model of Σ_0.

By the Compactness Theorem, Σ has a model X. Now give the kth colour to v_i in G if $R_k(c_i)$ holds in X. This is a legal colouring of G with r colours.

5.4 Use the notation of Exercise 5.2. By induction, we find that $(m, x) + n = (m+n, x)$; so two elements of the model lie in the same copy of $\mathbb{Z}$ if and only if they differ by a natural number. If c is greater then any natural number, then $mc - nc = (m - n)c$ is greater than any natural number if $m > n$; so all the elements nc for $n \in \mathbb{N}$ lie in different copies of $\mathbb{Z}$.

5.7 (a) Let the 2-element subsets of $\{0, 1, \ldots, 5\}$ be coloured red and blue in any manner. Consider the five pairs $\{x, 5\}$. At least three of them must have the same colour. Let $\{x, 5\}$, $\{y, 5\}$ and $\{z, 5\}$ be red. If any of the three pairs $\{x, y\}$, $\{y, z\}$, $\{z, x\}$ is red, say $\{x, y\}$, then we have a red triangle $\{x, y, 5\}$. Otherwise, $\{x, y, z\}$ is a blue triangle.

(b) Suppose that the colouring could be done so that $\{3, 4, 5\}$ was the only monochromatic triangle (say, it is red). Then, if x is any one of $0, 1, 2$, at least two of the edges $\{x, 3\}$, $\{x, 4\}$, $\{x, 5\}$ are blue (else another red triangle would be formed). So, for x and y in the set $\{0, 1, 2\}$, there exists $z \in \{3, 4, 5\}$ such that $\{x, z\}$ and $\{y, z\}$ are both blue. But then $\{x, y\}$ is red, to avoid creating a blue triangle. However, we have now created a red triangle $\{0, 1, 2\}$, contrary to assumption. So such a colouring is not possible.

This means that, in any colouring, there is a monochromatic triangle different from $\{3, 4, 5\}$. This triangle is large. So the conclusion of the Paris–Harrington Theorem holds.

6.3 (a) If $|A| = |B|$, then $|A| \leq |B \cup C|$. On the other hand,

$$|B \cup C| \leq |(B \times \{0\}) \cup (C \times \{1\})| = \max\{|B|, |C|\} = |A|.$$

Now the result follows from the Schröder–Bernstein Theorem.

(b) Clearly $|\mathcal{P}_n A| \leq |A^n| = |A|$. On the other hand, if $a_1, \ldots, a_{n-1}$ are distinct elements of A, then $|A| = |A \setminus \{a_1, \ldots, a_{n-1}\}|$, and there is an injection from $A \setminus \{a_1, \ldots, a_{n-1}\}$ to $\mathcal{P}_n A$ given by $x \mapsto \{a_1, \ldots, a_{n-1}, x\}$. So $|A| \leq |\mathcal{P}_n A|$. Again apply Schröder–Bernstein.

(c) We have

$$|\mathcal{P}_{\text{fin}} A| = 1 + |A| \cdot \omega = |A|.$$

6.4 The operation of symmetric difference on $\mathcal{P}_{\text{fin}} A$ makes this set a group. Now use the bijection from Exercise 6.3(c) to transfer this operation to A.

6.9 It is not very easy to think of one. The simplest example is the group C_{p^∞} whose elements are all the complex numbers which are p^m-th roots of unity, for some m, where p is a fixed prime. This group is infinite. It has a cyclic subgroup C_{p^m} of order p^m (consisting of all the p^m-th roots of unity) for each m. These are all the subgroups. For, if H is a subgroup containing one p^n-th root of unity, then H contains every p^m-th root of unity for all $m \leq n$. So if it has elements of arbitrarily large order, it must be the whole group.

The subgroups C_p^m form a chain whose union is the whole group. In a ring with identity, an ideal is proper if and only if it doesn't contain 1. There is no similar test for a subgroup of a group to be proper.

6.10 This is an example of using transfinite induction to construct something. Note that a cardinal is an ordinal with the property that all its sections have strictly smaller cardinality.

Following the hint, suppose that both X and $X^{(2)}$ have been well-ordered so that $X^{(2)}$ is isomorphic to a cardinal α. Now the construction is as follows. For each ordinal $\beta < \alpha$, we define a set S_β of triples. We start with $S_0 = \varnothing$.

Suppose that $\beta = s(\gamma)$. Consider the γ pair in the list, say $\{x_\gamma, y_\gamma\}$. If it is already contained in a triple in S_γ, then set $S_\beta = S_\gamma$. Otherwise, we seek a point z_γ such that neither $\{x_\gamma, z_\gamma\}$ nor $\{y_\gamma, z_\gamma\}$ are contained in triples of S_γ. Such points exist. For there are α points altogether, and at most γ triples in S_γ; and $|\gamma| < \alpha$ by assumption. So let z_γ be the first such point in the well-ordering, and set $S_\beta = S_\gamma \cup \{\{x_\gamma, y_\gamma, z_\gamma\}\}$.

If λ is a limit ordinal, let $S_\lambda = \bigcup_{\gamma < \lambda} S_\gamma$.

The final set S_α is a Steiner triple system: for we have checked every pair, and put it in a triple if it was not in one already; and no pair is ever put in more than one triple.

There is a shorter proof using Problem 6.3(c). If X is the set of finite non-empty subsets of A, and S the set of triples of sets whose symmetric difference is empty, then S is a Steiner triple system. But this argument does not generalise, whereas the previous one does.

6.11 Note that any set has only finitely many members, and for any finitely many natural numbers $m_1, \ldots, m_r$, there is a unique set containing just those members (namely $2^{m_1} + 2^{m_2} + \cdots + 2^{m_r}$).

Extension axiom: clear from the uniqueness just remarked.

Empty set axiom: 0 is the empty set.

Pair set axiom: Take 2^x if $x = y$, $2^x + 2^y$ otherwise. For example, $\{2, 3\} = 12$.

Union axiom: Take the binary expansion of n, say $n = 2^{m_1} + 2^{m_2} + \cdots + 2^{m_r}$. Then take the binary expansions of $m_1, \ldots, m_r$: let $p_1, \ldots, p_s$ be the distinct natural numbers which occur. Then

$$\bigcup n = 2^{p_1} + \cdots + 2^{p_s}.$$

For example, $\bigcup 24 = \bigcup \{4, 3\} = \{0, 1, 2\} = 7$.

Power set axiom: Let $n = 2^{m_1} + 2^{m_2} + \cdots + 2^{m_r}$, and take all sub-sums (including 0). Then 'assemble' these into a set by summing powers of 2. For example, $\mathcal{P} 3 = 2^3 + 2^2 + 2^1 + 2^0 = 15$.

Selection axiom, replacement axiom: trivial because these sets are finite, and we can construct any finite set!

Foundation axiom: If $x \in y$, then $x < 2^x \leq y$ (in the usual order). So, if Foundation fails, there is an infinite descending sequence of natural numbers, which is false.

Axiom of choice: Again trivial, since every finite set occurs.

6.12 Suppose that $f(x+y) = f(x) + f(y)$. Then by induction, $f(nx) = nf(x)$ for all positive integers n. Clearly this holds also for $n = 0$; and, if n is a negative integer, say $n = -m$, then $f(nx) + f(mx) = 0$, so $f(nx) = -f(mx) = -mf(x) = nf(x)$. So this equation holds for all integers n.

Now let $n = p/q$ be rational number, where p and q are integers. Then $qf(px/q) = f(px) = pf(x)$, so $f(px/q) = (p/q)f(x)$. So the equation holds for all rational numbers.

If r is a real number, then $r = \lim_{n\to\infty} q_n$ is the limit of a sequence of rational numbers; and we have

$$f(rx) = \lim_{n\to\infty} f(q_n x) = \lim_{n\to\infty} q_n f(x) = rf(x),$$

where we have used the continuity of f in the first step. So the equation holds for all real numbers r. In particular,

$$f(x) = xf(1) = cx,$$

where $c = f(1)$.

We can regard $\mathbb{R}$ as a vector space over $\mathbb{Q}$, by 'forgetting' the multiplication. Choose a basis $(e_i : i \in I)$ for this vector space, where we may take $e_0 = 1$, $e_1 = \sqrt{2}$. Now, for any assignment of values $f(e_i)$ for $i \in I$, there is a unique linear function f taking these values. We set $f(e_0) = 1$ and $f(e_1) = 0$. Then $f(x+y) = f(x) + f(y)$ (this is half of the assertion that f is a linear function on the vector space), but $f(\sqrt{2}) \neq \sqrt{2}f(1)$. By the first part of the question, f must be discontinuous.

6.13 If $x \in \bigcup x$, then $x \in y$ for some $y \in x$, and we have an infinite descending chain

$$\ldots \in y \in x \in y \in x.$$

Every set x satisfies $x \in \mathcal{P}\,x$ (this just means $x \subseteq x$).

6.15 We saw that any model M of the axioms for the successor function has the form $\mathbb{N} \cup (\mathbb{Z} \times X)$ for some set X. Now $|M| = \omega + |X| \cdot \omega$. Moreover, two models are isomorphic if and only if the sets X have the same cardinality. If X is uncountable, then $|M| = |X|$, so the cardinality of M determines that of X. However, any finite or countable set X gives a countable model M, so there are countably many non-isomorphic countable models (with $|X| = 0, 1, 2, \ldots, \omega$). So the theory is α-categorical for all uncountable α but not ω-categorical.

7.1 We take $O = I$, with $1_i = i$ for all $i \in I$. The functions dom and cod are implicitly defined by the second condition. Now verify the axioms given earlier.

7.2 Take the proof for sets and set functions and write it out, using the fact that a homomorphism $f : G \to H$ between two groups is an isomorphism if there is a homomorphism $f' : H \to G$ such that ff' and $f'f$ are both identity homomorphisms.

7.4 The first part is Exercise 1.8. For the converse, let $G = C_2 = \{1, a\}$ and $H = C_4 = \{1, b, b^2, b^3\}$. There are two homomorphisms from G to H, namely f_0 (which maps everything to 1) and f_1 (which maps 1 to 1 and a to b^2); and two homomorphisms from H to G, namely g_0 (which maps everything to 1) and g_1 (which maps 1 and b^2 to 1, and b and b^3 to a). Now $f_1 g_1$ maps everything in C_2 to the identity, while $g_1 f_1$ maps x to x^2 for all $x \in C_4$. So none of these homomorphisms have left or right inverses. However, f_1 is one-to-one, and g_1 is onto.

7.8 (a) Since (u_1, e_1) is an equaliser, there is a morphism $h : u_2 \to u_1$ such that $he = e'$. Similarly, there is a map $h' : u_1 \to u_2$ such that $h'e' = e$. Now $(h'h)f_1 = (h'h)f_2$; the uniqueness clause in the definition of an equaliser shows that $h'h = 1_{u_1}$. Similarly, $hh' = 1_{u_2}$. So h and h' are inverse isomorphisms between u_1 and u_2.

(b) In the category of sets, let $u = \{a \in x : af_1 = af_2\}$, and let e be the embedding of u into x. Then (u, e) is an equaliser of f_1 and f_2. This construction explains the name: u is the set on which the maps are equal.

(c) A *coequaliser* of f_1 and f_2 is a pair (v, d), where v is an object and $d : y \to v$ a morphism, satisfying

- $f_1 d = f_2 d$;
- for any morphism $g : y \to z$ such that $f_1 g = f_2 g$, there is a unique morphism $h : v \to z$ such that $dh = g$.

Now, if (v_1, d_1) and (v_2, d_2) are coequalisers, then v_1 and v_2 are isomorphic; the proof is dual to the proof of (a) above.

The construction of coequalisers in the category Set is a little harder. We need a set v and function $d : y \to v$ such that $a(f_1 d) = a(f_2 d)$ for all $a \in x$. Define a relation R on y by

$$R = \{(af_1, af_2) : a \in x\}.$$

Now let S be the equivalence relation generated by R (the reflexive, symmetric and transitive closure); set $v = y/S$ and let d be the projection map from y to v.

7.9 If the universes form a set, then there is a universe containing it.

7.12 (a) The free group on x can be represented as the set of words in symbols $\{a^{+1}, a^{-1} : a \in x\}$; the group operation is juxtaposition followed by all possible cancellation. The functor FG maps a set x to the set of all such words, but ignores the multiplication.

(b) The natural transformation T in question should map any set x to the natural inclusion map $i_x : a \mapsto a^{+1}$ from x to xFG. Given a function $f : x \to y$ (a morphism in the category **Set**), there is an induced map fFG from xFG to yFG which replaces each symbol $a^{\pm 1}$ with $(af)^{\pm 1}$ and then performs the cancellation. To show that T is a natural transformation, it is required to show that the square

$$
\begin{array}{ccc}
x & \xrightarrow{\ f\ } & y \\
i_x \downarrow & & \downarrow i_y \\
xFG & \xrightarrow{\ fFG\ } & yFG
\end{array}
$$

is commutative.

7.14 There is no natural way in which a homomorphism $f : G \to H$ induces a homomorphism from $Z(G)$ to $Z(H)$. For example, if $G = C_3$ (the cyclic group) and $H = S_3$ (the symmetric group), then the inclusion map from G to H is a monomorphism; but $Z(G) = G$ and $Z(H) = \{1\}$, so the map induced by the inclusion does not map $Z(G)$ to $Z(H)$, and the only homomorphism which does so is trivial.

7.15 (a) Let (X, P) be a preordered set: that is, X is a set and P a preorder on X. Let $O = X$ and $M = P$ (a set of ordered pairs). For each $(x, y) \in P$, let $\mathrm{dom}((x, y)) = x$ and $\mathrm{cod}((x, y)) = y$. Moreover, let $1_x = (x, x)$. Then we have a category. (Most of the verification is obvious. To check closure under composition, suppose that $\mathrm{cod}(f) = \mathrm{dom}(g)$. Then, say, $f = (x, y)$ and $g = (y, z)$. By assumption, $(x, z) \in P$, and this is the only possible choice for the composition of f and g.)

(b) Obviously such a category has at most one morphism with any given domain and codomain. Conversely, let $C = (O, M)$ be a category with this property. Let

$$P = \{(x, y) \in O \times O : (\exists f \in M) \text{ with } \mathrm{dom}(f) = x \text{ and } \mathrm{cod}(f) = y\}.$$

We claim that P is a preorder on O:

- It is reflexive, since for any $x \in O$ the morphism 1_x has domain and codomain x.

- It is transitive, since if $(x, y), (y, z) \in P$, with say $f : x \to y$ and $g : y \to z$, then the composition fg is defined and has domain x and codomain z, so that $(x, z) \in P$.

References

1. William A. Adkins and Steven H. Weintraub, *Algebra: An Approach via Module Theory*, Springer, New York, 1992.
2. Kenneth Appel and Wolfgang Haken, Every planar map is four colourable, *Bull. Amer. Math. Soc.* **82** (1976), 711–712.
3. Jon Barwise (ed.), *Handbook of Mathematical Logic*, North-Holland, Amsterdam, 1977.
4. Jon Barwise and Lawrence Moss, *Vicious Circles: On the Mathematics of Non-Wellfounded Phenomena*, CSLI, Stanford, 1996.
5. Errett Bishop and Douglas S. Bridges, *Constructive Analysis*, Springer-Verlag, Berlin, 1985.
6. Norman L. Biggs, *Discrete Mathematics*, Oxford University Press, Oxford, 1989.
7. T. S. Blyth and E. F. Robertson, *Basic Linear Algebra*, Springer Undergraduate Mathematics Series, Springer, London, 1998.
8. Felix E. Browder (ed.), *Mathematical Developments arising from Hilbert Problems*, Proceedings of Symposia in Pure Mathematics **28**, American Mathematical Society, Providence, RI, 1976.
9. Peter J. Cameron, *Introduction to Algebra*, Oxford University Press, Oxford, 1998.
10. John N. Crossley *et al.*, *What is Mathematical Logic?*, Oxford University Press, Oxford, 1972.
11. Nigel J. Cutland, *Computability: An Introduction to Recursive Function Theory*, Cambridge University Press, Cambridge, 1980.
12. Tobias Dantzig, *Number: The Language of Science*, Macmillan, New York, 1930.
13. Stanislas Dehaene, *The Number Sense: How the Mind Creates Mathematics*, Oxford University Press, New York, 1997.
14. Keith J. Devlin, *The Joy of Sets*, Springer, Berlin, 1979.
15. Herbert B. Enderton, *Elements of Set Theory*, Academic Press, New York, 1977.
16. Ronald L. Graham, Bruce L. Rothschild and Joel H. Spencer, *Ramsey Theory* (2nd edition), Wiley, New York, 1990.
17. Jacques Hadamard, *The Psychology of Invention in the Mathematical Field*, Princeton University Press, Princeton, 1945; reprinted as *The Mathematician's Mind*, Princeton University Press, Princeton, 1968.
18. A. G. Hamilton, *Logic for Mathematicians*, Cambridge University Press, Cambridge, 1978.
19. A. Heyting, *Intuitionism: An Introduction*, North-Holland, Amsterdam, 1966.
20. Russell Hoban, *Pilgermann*, Pan, London, 1984.

21. Wilfrid Hodges, *A Shorter Model Theory*, Cambridge University Press, Cambridge, 1997.

22. Douglas R. Hofstadter, *Gödel, Escher, Bach: An Eternal Golden Braid*, Basic Books, New York, 1979.

23. Georges Ifrah, *From One to Zero: A Universal History of Numbers*, Penguin, London, 1987.

24. Julian Jaynes, *The Origin of Consciousness in the Breakdown of the Bicameral Mind*, Houghton Mifflin, New York, 1976.

25. David L. Johnson, *Presentations of Groups*, Cambridge University Press, Cambridge, 1990.

26. David L. Johnson, *Elements of Logic via Numbers and Sets*, Springer Undergraduate Mathematics Series, Springer, London, 1998.

27. Peter T. Johnstone, *Topos Theory*, Academic Press, London, 1977.

28. Richard Kaye, *Models of Peano Arithmetic*, Oxford University Press, Oxford, 1991.

29. Richard Kaye and Dugald Macpherson (ed.), *Automorphisms of First-Order Structures*, Oxford University Press, Oxford, 1994.

30. David Knowles, *The Evolution of Mediaeval Thought*, Longman, London, 1962.

31. Donald E. Knuth, *Surreal Numbers*, Addison–Wesley, Reading, MA, 1974.

32. Imre Lakatos, *Proofs and Refutations: The Logic of Mathematical Discovery*, Cambridge University Press, Cambridge, 1976.

33. C. W. H. Lam, S. Swiercz and L. Thiel, The nonexistence of finite projective planes of order 10, *Canad. J. Math.* **41** (1989), 1117–1123.

34. J. E. Littlewood, *A Mathematician's Miscellany*, Methuen, London, 1953; reprinted in *Littlewood's Miscellany* (ed. Béla Bollobás), Cambridge University Press, Cambridge, 1986.

35. Saunders MacLane, *Categories for the Working Mathematician*, Springer, Berlin, 1971.

36. Saunders MacLane, *Mathematics: Form and Function*, Springer, New York, 1986.

37. Colin McLarty, *Elementary Categories, Elementary Toposes*, Oxford University Press, Oxford, 1995.

38. John von Neumann and Oskar Morgenstern, *Theory of Games and Economic Behavior*, Princeton University Press, Princeton, 1944.

39. Benjamin C. Pierce, *Basic Category Theory for Computer Scientists*, MIT Press, Boston, 1991.

40. Henri Poincaré, *Science and Hypothesis*, Walter Scott, 1905; reprinted Dover Publications, New York, 1952.

41. Bertrand Russell, *History of Western Philosophy*, George Allen and Unwin, London, 1961.

42. Geoff Smith, *Introductory Mathematics: Algebra and Analysis*, Springer Undergraduate Mathematics Series, Springer, London, 1998.

43. Raymond Smullyan, *What is the name of this book? : The riddle of Dracula and other logical puzzles*, Prentice-Hall, Englewood Cliffs, NJ, 1978.

44. B. L. van der Waerden, How the proof of Baudet's conjecture was found, pp. 251–260 in *Studies in Pure Mathematics* (ed. L. Mirsky), Academic Press, London, 1971.

45. Stan Wagon, *The Banach–Tarski Paradox*, Cambridge University Press, Cambridge, 1985.

46. D. A. R. Wallace, *Groups, Rings and Fields*, Springer Undergraduate Mathematics Series, Springer, London, 1998.

Index

action 33
Aczel, P. 134
addition
 – of cardinals, 126
 – of ordinals, 51
Adkins, W. A. 158
aleph notation 125
algebraic numbers 35
algebraically closed 31
alphabet 56
Anti-Foundation Axiom 134
antisymmetric 11
Appel, K. vi, 159
Arisotle 159
arity 81
atomic formula 82
automorphism 100
automorphism group 100
axiom 56
Axiom of Choice 11, 16, 32, 114
Axiom of Determinacy 134
Axiom of Infinity 114

back-and-forth 139
Banach, S. 122
Banach–Tarski paradox 122
Barwise, J. v, 158
basis 121
Biggs, N. L. 158
bijective 9
binary 80
binary relation 10
Bishop, E. 157
Blyth, T. S. vii
Boolean algebra 70

Boolean ring 73, 134
bound variable 82
Bridges, D. S. 157
Burali-Forti paradox 44

Cameron, N. J. viii
Cameron, P. J. 158
cancellation laws 28
Cantor's Theorem 19, 135
Cantor, G. 2, 19
cardinal 124
cardinal arithmetic 126
cartesian product 8, 11
category
 – functor, 146, 154
 – locally small, 147
 – small, 146
 – well-powered, 147
chain 118
characteristic function 128
choice function 11
class 136
closed term 92
coequalizer 154
cofinal 132
cofinality 132
Cohen, P. J. 129
colouring 67, 76
commutative square 150, 152
Compactness Theorem 67, 95
comparable 118
composition 10
conjugacy class 33
conjugation 33
connective 58, 81

consistent   64
constant symbol   81
constructivism   157
continuum   129
Continuum Hypothesis   129
contradiction   61
converse   34
coproduct   146
countable   24
countably additive   122
countably categorical   99
Crossley, J. N.   134
Cutland, N. J.   158

Darwinism   133
decision procedure   56, 67
Dedekind cut   30
Dedekind finite   21
Dedekind infinite   21
Deduction Theorem   63, 87
Dehaene, S.   156
Descartes, R.   8
Devlin, K. J.   158
difference   7
disjoint   6
disjunctive normal form   72
Dugan, M.   31

edge   139
Einstein, A.   133
empty set   5
Empty Set Axiom   114
Enderton, H. B.   158
epic   144
Epimenides   2
epimorphism   144
∈-model   132, 147
equalizer   153
equals sign   81
equivalence   101
equivalence relation   12
exponentiation
– of cardinals, 126
– of ordinals, 51
Extension Axiom   114

family of sets   10
field   30
field of fractions   30
finite   21
finite Ramsey Theorem   107
first-order logic   vi, 79
forcing   129
forgetful functor   149

formal system   56
formula   56, 81, 82
– atomic, 82
– well-formed, 56, 81
Foundation Axiom   114, 134
Four-Colour Theorem   68, 97, 159
free variable   82
Frege, G.   2
function   9
– n-ary, 80
function symbol   81
functor   148
– forgetful, 149
functor category   146, 154

general relativity   133
Generalization   86
Generalized Continuum Hypothesis   130
Gödel numbering   103
Gödel's Incompleteness Theorem   106
Gödel's Second Incompleteness Theorem   4, 110
Gödel, K.   4, 103, 129
Goldbach's Conjecture   102
Graham, R. L.   109
graph   76, 139
greatest element   15
Grothendieck's Axiom   148
Grothendieck, A.   147

Hadamard, J.   157
Haken, W.   vi, 159
Hamel basis   139
Hamilton, A. G.   158
Harrington, L.   107
Hegel, G. W. F.   157
Heraclitus   157
Heyting, A.   157
Hilbert, D.   129
Hoban, R.   155
Hodges, W. A.   134, 158
Hofstadter, D. R.   56, 103
hom-functor   153
Hypergame   3

identity   10
Ifrah, G.   16
image   13
inaccessible cardinal   130
inconsistent   64
index set   10
induction   38, 45
infinite   21

infinite Ramsey Theorem   108
Infinity Axiom   134
injective   9
integers   29
intersection   6, 10
inverse   10, 145
irreflexive   11
isomorphism   15, 98, 145

Jaynes, J.   v
Jensen's Diamond   134
Johnson, D. L.   vii
Johnstone, P.   158
joined   139
Joshu's Mu   58

Kaye, R. W.   158
kernel   13
Knowles, D.   2
Knuth, D. E.   157
koan   58
Kronecker, L.   28

$\mathcal{L}$-structure   83
Lam, C. W. H.   159
Lawvere, F. W.   148
least element   14
Lebesgue measure   121
left coset   33
Leibniz, G. W.   5, 159
lexicographic product   50
Liar Paradox   2
limit ordinal   46
Littlewood, J. E.   3, 9, 157
locally small category   147
logic
– first-order, vi, 79
– propositional, 58
– second-order, 80
logical consequence   61
logically equivalent   70
logically valid   88
Löwenheim–Skolem Theorem   95

MacLane's Axiom   148
MacLane, S.   v, 109, 141, 158
Macpherson, H. D.   158
Martin's Axiom   134
maximal element   15
maximal ideal   120
McLarty, C.   158
measurable set   121
membership   4
metatheorem   56

minimal element   14
model   84
– non-standard, 103
Modus Ponens   62, 86
monic   144
monomorphism   144
Morgenstern, O.   113
Moss, L.   v, 158
MU-system   56
multiplication
– of cardinals, 126
– of ordinals, 51
Multiplicative Axiom   32

n-ary function   80
n-ary relation   80
natural numbers   15, 40, 51
von Neumann, J.   113
Newson, M. W.   129
non-Euclidean geometry   133
non-measurable set   122
non-standard model   103

oligomorphic   100
ω-categorical   99
ω-consistency   105
one-to-one   9
onto   9
orbit   33, 100
order   13
– non-strict, 13
– partial, 13
– strict, 13
– total, 14
ordered pair   7
ordered set   14
ordered sum   50
ordinal   40
ordinal product   51
ordinal sum   51

Pair Set Axiom   114
Paris, J.   107
Paris–Harrington Theorem   108
parsing   59
parsing tree   59
partition   12
Peano arithmetic   103
Peano finite   21
Peano infinite   21
Pierce, B. C.   158
plane map   67
Plato   2
Poincaré, H.   157

power set   6
Power Set Axiom   114
Principle of Extension   5
Principle of Induction   39, 103
Principle of the Supremum   30
product   145
proof   56
proper class   136, 137
proper subset   6
Propositional Compactness   134
propositional logic   58
Propositional Soundness and Completeness Theorem   64
propositional variable   58

quantifier   81

Ramsey's Theorem   107, 108
recursion theory   158
Reed–Muller code   77
reflexive   11
regular cardinal   132
relation   10
– binary, 10
– n-ary, 80
relation symbol   81
Replacement Axiom   114
Riemann, B.   133
right multiplication   33
ring   29
Robertson, E. F.   vii
Rothschild, B. L.   109
rule of inference   56
Russell's Paradox   2
Russell, B.   2, 31, 155

satisfiable   67
Schröder–Bernstein Theorem   17, 126
scope   82
section   40
Selection Axiom   114
semantics   59
sentence   82
singular cardinal   132
Skolem Paradox   134
small category   146
Smith, G.   vii
Smullyan, R.   91
Soundness and Completeness Theorem
   89
Spencer, J. H.   109
Stackpool, W.   31
Steiner triple system   138
subformula   82

subset   5
successor   102
successor ordinal   46
surjective   9
symmetric   11

Tarski's Theorem   107
Tarski, A.   122
tautology   61
term   81
theorem   56
Theorem of Engeler, Ryll-Nardzewski
   and Svenonius   100
theory   84, 101
Thierry of Chartres   2
topology   80
total order   14
transcendental numbers   35
transfinite induction   45
transitive   11
transitive set   132
trichotomy   14
Truss, J. K.   124
Typographic Number Theory   103

unary   80
union   6, 10
Union Axiom   114
universe   147
upper bound   118
Upward Löwenheim–Skolem Theorem
   97, 130
urelement   5

valuation   60, 83
variable   81
vertex   139

van der Waerden, B. L.   157
Wagon, S.   122
Wallace, D. A. R.   vii
Weintraub, S. H.   158
well-formed formula   56, 81
well-order   38
well-ordered set   38
Well-Ordering Principle   118, 134
well-powered category   147
William of Ockham   159
witness   90

Zen Buddhism   58
Zermelo's hierarchy   47
Zermelo–Fraenkel axioms   114
zero   46
Zorn's Lemma   118